Basic Motorsport Engineering

Units for Study at Level 2

D1354573

Basic Motorsport Engineering

Units for Study at Level 2

Andrew Livesey

ELSEVIER

AMSTERDAM • BOSTON • HEIDELBERG • LONDON • NEW YORK
OXFORD • PARIS • SAN DIEGO • SAN FRANCISCO
SINGAPORE • SYDNEY • TOKYO

Butterworth-Heinemann is an imprint of Elsevier

Butterworth-Heinemann is an imprint of Elsevier
The Boulevard, Langford Lane, Oxford OX5 1GB, UK
30 Corporate Road, Burlington, MA 01803

First edition 2011

Notices
Knowledge and best practice in this field are constantly changing. As new research and experience broaden our understanding, changes in research methods, professional practices, or medical treatment may become necessary.

Practitioners and researchers must always rely on their own experience and knowledge in evaluating and using any information, methods, compounds, or experiments described herein. In using such information or methods they should be mindful of their own safety and the safety of others, including parties for whom they have a professional responsibility.

To the fullest extent of the law, neither the Publisher nor the authors, contributors, or editors, assume any liability for any injury and/or damage to persons or property as a matter of products liability, negligence or otherwise, or from any use or operation of any methods, products, instructions, or ideas contained in the material herein.

British Library Cataloguing in Publication Data
A catalogue record for this book is available from the British library

Library of Congress Cataloguing in Publication Data
A catalogue record for this book is available from the library of Congress

ISBN: 978-0-75-068909-0

For information on all Butterworth-Heinemann Publications
visit our website at www.elsevierdirect.com

Printed in Great Britain

11 10 9 8 7 6 5 4 3 2 1

Working together to grow
libraries in developing countries

www.elsevier.com | www.bookaid.org | www.sabre.org

ELSEVIER   BOOK AID International   Sabre Foundation

CONTENTS

This is the first book I know of to be specifically written for the formal training of motorsport engineers.

In the press you will read about the major racing teams in Formula 1, ALMS and NASCAR. In addition to those classes, there are thousands of other more affordable forms of racing, which take place on permanent and temporary circuits near every major town and city in the UK and America.

Motorsport Engineering is an excellent subject to study as it teaches you to investigate and analyse situations, as well as learning a practical skill and investigating current and future technology.

I hope that all of you who read this book and take up a career in Motorsport Engineering will have as much fun as I have had over the past 50 years in a career which led me from the UK, across Europe, Africa, New Zealand, South America and the USA.

Brian Redman
www.gorace.com

If you are looking for a career in motorsport and you want a course at Level 2, this is the book for you; or maybe you have started on a more advanced course, but need a detailed explanation of some of the motorsport stuff that you don't want to ask about — this is your book too — anybody keen on cars and motorsport will find some interesting material in this book.

I ran the motorsport and the high performance car courses for Oxford Brookes University based at Brooklands; the latter course was used to train many of the McLaren Group staff. Ex-students can be found in most of the major teams. However, it is worth noting that the bulk of motorsport takes place at grass-roots level — in other words, club motorsport.

This book covers the essential knowledge for most motorsport courses at Level 2 — IMI, EAL and BTEC.

Any student interested in motorsport is advised to get some work experience. In the first instance you'll be brewing the tea and passing the spanners before you will be allowed to work on your own, so read this book in your spare time and remember that motorsport is a lifestyle, not a job, have fun and look out for me at races and other events; I'm usually either in a green MG, or on a red Honda.

Andrew Livesey MA MIMechE AAE FIMI
Herne Bay, Kent
E-mail: Andrew@BrooklandsGreen.com

ix

Thanks to the following for their help and support with this book; either with information, direct or indirect support, which enabled the writing process and my involvement in motorsport to continue.

Adrian Slaughter
Alan Flavell
Alex Heaton
Andy Stevens
Automotive Skills
Bernie Ecclestone
Bill Sisley
Brian Marshall
Brian Redmond
Brooklands Museum
BrooklandsGreen.Com
Buckmore Park Kart Circuit
Bucks New University
Canterbury College
Carol Dorman
Chris Pulman
Claire Kocks
Daniel Johnson
Evers Pearce
Express Garage, Herne Bay
Frank Groom
Helix Autosport Ltd
Ian Goodwin
Institute of the Motor Industry
Jack Sealey Ltd
Jenson Button
John Hunt
John Ryan
John Surtees
Kevin Pilcher
Laser Tools
Linda Teuton
McLaren Cars
McLaren Racing
Mick Ellender
Mike Farnworth

ACKNOWLEDGEMENTS

Mike Shaw
Motorsport Academy
Motorsport Institute
Mr and Mrs Presgraves (Snr)
Nick Sansom
North West Kent College
Oxford Brookes University
Paul Presgraves
Paulio Racing
Performance Racing Industry
Proton Cars (UK) Ltd
R T Quaife Engineering Ltd
Rallyspeed
Richard Davies
Seyed Edalatpour
Simon Hammond
Staffordshire University
Steven Swaffer
Terry Ormerod
The late James Hunt
Tool Connection
University for the Creative Arts
University of Brighton
University of Greenwich
Winston Sewell

The abbreviations are generally defined by being written in full when the relevant technical term is first used in the book. In a very small number of cases an abbreviation may be used for two separate purposes, usually because the general concept is the same, but the use of a superscript or subscript would be unnecessarily cumbersome; in these cases the definition should be clear from the context. The units used are those of the internationally accepted *System International* (SI). However, because of the large American participation in motorsport, and the desire to retain the well-known imperial system of units by UK motorsport enthusiasts, where appropriate, the imperial equivalents of SI units are given. Therefore, the following is intended to be useful for reference only and is neither exhaustive nor definitive.

A Ampere
ABDC after bottom dead centre
ABS anti-lock braking system; or acrylonitrile butadiene styrene (a plastic)
AC alternating current
AED automatic enrichment device
AF across flats – bolt head size
ATC automatic temperature control
bar atmospheric pressure
BBDC before bottom dead centre
BDC bottom dead centre
BS British Standard
BSI British Standards Institute
BTDC before top dead centre
C Celsius; or centigrade
CG centre of gravity
CI compression ignition
cb contact breaker
cm centimetre
cm3 cubic centimeters – capacity; engine capacity also called cc. 1000 cc is 1 litre
CO carbon monoxide
CO$_2$ carbon dioxide
COSHH Control of Substances Hazardous to Health (Regulations)
CR compression ratio
D diameter
d distance
DC direct current
deg degree (angle or temperature)
dia diameter
DTI dial test indicator
EC European Community
ECU Electronic Control Unit
EF electronic fuel injection
EPA Environmental Protection Act; or Environmental Protection Agency
EU European Union

F Fahrenheit

ft foot

ft/min feet per minute

FWD front-wheel drive

g gravity; or gram

gal gallon (USA gallon is 0.8 of UK gallon)

GRP glass reinforced plastic (glass fibre)

HASAWA Health and Safety at Work Act

HGV heavy goods vehicle (used also to mean LGV — large goods vehicle)

HP horse power (CV in French, PS in German)

HSE Health and Safety Executive; also health, safety and environment

HT high tension (ignition)

I inertia

ICE in-car entertainment

ID internal diameter

IFS independent front suspension

IMechE Institution of Mechanical Engineers

IMI Institute of the Motor Industry

in3 cubic inches — measure of capacity; also cu in. Often called 'cubes' — 61 cu in is approximately 1 litre

IRS independent rear suspension

ISO International Standards Organization

Kph kilometres per hour

KW kerb weight

l length

L wheelbase

LH left hand

LHD left-hand drive

LHThd left-hand thread

LPG liquid petroleum gas

LT low tension (12 volt)

lumen light energy radiated per second per unit solid angle by a uniform point source of 1 candela intensity

lux unit of illumination equal to 1 lumen/m^2

M mass

MAX maximum

MDRV mass driveable vehicle

MI Motorsport Institute

MIA Motorsport Industry Association

MIG metal inert gas (welding)

MIN minimum

MOT Ministry of Transport; also called DfT — Department for Transport, and other terms depending on the flavour of the government, such as the Department of the Environment, Transport and the Regions (DETR), not to be confused with DOT which is the American equivalent

mpg miles per gallon

mph miles per hour

N Newton

Nm Newton metre (torque)

No number

OD outside diameter

OHC overhead cam

OHV overhead valve

OL overall length

OW overall width

P power, pressure or effort

Part no part number

PPE Personal Protective Equipment

Psi pounds per square inch
PSV public service vehicle (also used to mean PCV – public-carrying vehicle, in other words a bus)
pt pint (UK 20 fluid ounces, USA 16 fluid ounces)
PVA polyvinyl acetate
PVC polyvinyl chloride
r radius
R reaction
Ref reference
RD relative density
RH right hand
RHD right-hand drive
rpm revolutions per minute; also RPM and rev/min
RTA Road Traffic Act
RWD rear-wheel drive
SAE Society of Automotive Engineers (USA)
SI spark ignition
SIPS side impact prevention system
SRS supplementary restraint systems
std standard
STP standard temperature and pressure
SW switch
TDC top dead centre
V volt
VIN vehicle identification number
VOCs volatile organic compounds
W weight
w width
WB wheel base

Superscripts and subscripts are used to differentiate specific concepts.

SI Units

cm centimetre
K Kelvin (absolute temperature)
kg kilogram (approx. 2.25 lb)
km kilometre (approx. 0.625 mile or 1 mile is approx. 1.6 km)
kPa kilopascal (100 kPa is approx. 15 psi, that is atmospheric pressure of 1 bar)
kV kilovolt
kW kilowatt
l litre (approx. 1.7 pint)
l/100 km litres per 100 kilometres (fuel consumption)
m metre (approx. 39 inches)
mg milligram
ml millilitre
mm millimetre (1 inch is approx. 25 mm)
N Newton (unit of force)
Pa Pascal
μg microgram

Imperial Units

ft foot (12 inches)
hp horsepower (33 000 ftlb/minute; approx. 746 Watt)
in inch (approx. 25.4 mm)
lb/in^2 pressure, sometimes written psi
lbft torque (10 lbft is approx. 13.5 Nm)

CHAPTER 1

Body and Chassis

KEY POINTS

- Most vehicles are front-wheel drive layout; alternative layouts are: conventional, mid-engine, and rear engine.
- The chassis is load bearing and must be free of corrosion.
- Jacking points and seat belt mounting points are specially reinforced.
- Air bags must only be handled following a special procedure.
- SIPS and crumple zones are sometimes added to give extra passenger protection.

On most popular cars the body and the chassis are one and the same. Trucks and buses, however, are likely to have the body and the chassis as separate components. Motorsport vehicles may use either arrangement. The chassis is the part to which the engine, gearbox, suspension and other components are attached. The **body** is the covering for the components, the passengers and the load. The **chassis** is load bearing, being made from strong steel. The body does not carry a load and may be made from aluminium alloy, or some form of plastic, as well as the more usual steel.

1. VEHICLE LAYOUT

By vehicle layout we mean the position of the engine and gearbox on the chassis in relation to the driving wheels.

Front-wheel drive (FWD) is the most common; the engine and the gearbox are mounted at the front of the car and short driveshafts take the power to the front wheels.

Conventional layout means that the engine is mounted at the front with a gearbox behind it and a propeller shaft takes the power to a rear axle so that the rear wheels are driven.

Rear-engined vehicles drive the rear wheels from a rear-mounted engine (RWD).

1

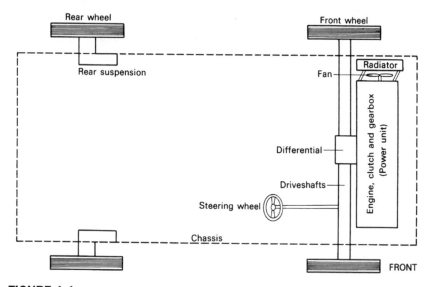

FIGURE 1.1
Front engine front wheel drive (FWD)

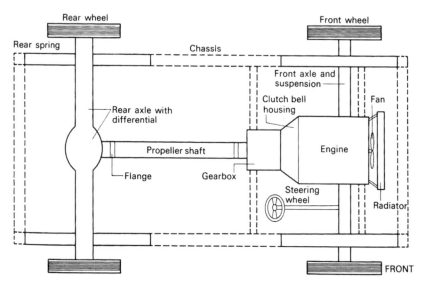

FIGURE 1.2
Front engine rear wheel drive

Increasingly **mid-engine** set-ups are becoming more popular in high performance cars, where the engine and the gearbox are mounted in the middle of the vehicle and the rear wheels are driven.

For off-road use **four-wheel drive** gives better grip. This can be with front or mid-engined layouts.

Nomenclature

4 × 4 means that the vehicle has a total of four wheels and that it is driven by all the four wheels. 4×4 vehicles may also be called all-wheel drive (AWD), off-road, or all-terrain vehicle.

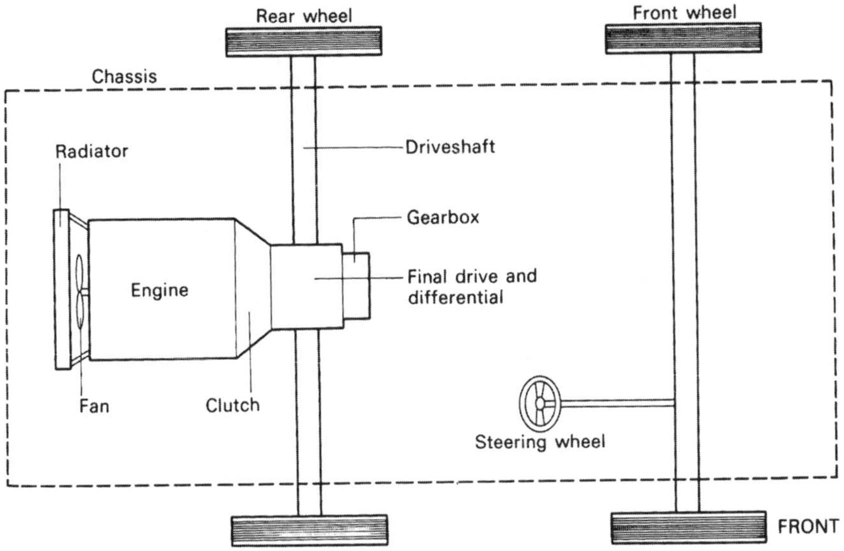

FIGURE 1.3
Rear engine rear wheel drive (RWD)

2. ADVANTAGES AND DISADVANTAGES OF DIFFERENT VEHICLE LAYOUTS

Each type of layout has certain advantages and disadvantages when compared to the others, see Table 1.1.

3. CHASSIS

The chassis is the **load bearing** part of the vehicle. That is to say it carries the weight of the **load** and the **passengers,** and locates the engine, transmission,

TABLE 1.1

Type of Layout	Advantages	Disadvantages
FWD	• good traction • maximum passenger space	• noise from transmission • complicated driveshafts
Conventional	• easy to repair • high level of safety	• passenger space limited by drive tunnel
Mid-engine	• best weight distribution • good for 4×4 transmission	• limited passenger space • difficult to access engine
RWD	• engine noise behind the driver • good traction	• long control cable/rods needed • luggage space limited — between front wheels and location of fuel tank is difficult

3

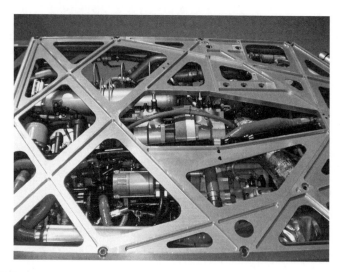

FIGURE 1.4
CNC machined chassis of the JCB Diesel record holder car

steering and suspension. On most popular cars the chassis and the body are one and the same; but on specialized cars and goods vehicles separate chassis are used. There are three main types of chassis: these are **ladder chassis**, **cruciform chassis** and **backbone chassis**.

▌ RACER NOTE

If you look underneath a popular car you will see square sections, like little box girders, connecting the main suspension and transmission parts. These are called the chassis sections; although the vehicle does not have a separate chassis these give extra strength. It is important that the chassis sections are in good condition for the car to be safe.

Ladder chassis — it is called this because of its shape: it looks vaguely like a builder's ladder. It has two side rails connected by cross-members. Ladder

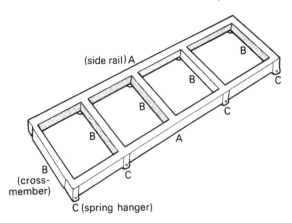

FIGURE 1.5
Ladder chassis

4

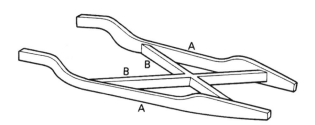

FIGURE 1.6
Cruciform chassis

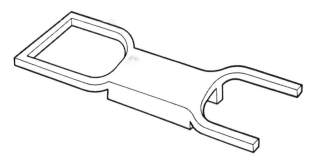

FIGURE 1.7
Backbone chassis

chassis are used on trucks and buses as well as vehicles like Land Rovers and some kit cars. The rails which run lengthwise are called **longitudinal** members; those which go across the vehicle are **transverse** members. Generally the suspension is attached to the longitudinal rails and the engine will sit between these rails. The gearbox tail housing is attached to a transverse member.

Cruciform chassis – it is cross-shaped in the middle to give resistance to twisting. This type of chassis is used on some old rare sports cars such as Lea-Francis.

Backbone chassis – it looks roughly like a person's skeleton – a backbone with arms and legs. Lotus and Mazda use this design on their small sports cars (Elise and MX-5); the propeller shaft can fit through the middle of the hollow backbone section.

4. INTEGRAL CONSTRUCTION

Integral construction is also known as **unitary construction** or **monocoque**. This is when the chassis and the body are made as one integral unit, that is, as one piece from parts welded together, not as a separate body and chassis. These are often referred to as body shells. The floor, the sills, the roof and the quarter panels are all spot-welded together to form an assembly to which the engine and the running gear are attached. The integral body/chassis is much lighter that using separate components; it is also very strong, especially in resistance to twisting, which makes the car feel good to drive.

5

Tubs

Race car body and chassis units are often constructed as tubs. That is, they are made from composite material — usually carbon fibre is laid over honeycomb section board — to form an integral unit. The suspension and engine mounting points are built into the tub.

Nomenclature
Integral, unitary and monocoque, in relation to body/chassis means that it is made as one piece — that is the body and chassis is a two-in-one. The word 'tub' is an abbreviation for bathtub, in other words the open racing car tub resembles a bathtub, or a hot tub.

Birdcages

Birdcage construction is where the chassis is made from tubular steel so that the engine, transmission and drive line sit inside it. Then a non-stressed bodyshell is fitted over the birdcage to cover it in.

Nomenclature
Although tubular usually refers to a round section, it can be round or square, as long as it is hollow.

5. CHASSIS SECTIONS

Different shapes of chassis sections are used for different purposes. The **round section** is the strongest closely followed by the **square section**. The square section has the advantage of being easy to screw or weld components to. The round section is used for bicycle and motorcycle frames, and on specialist racing cars like Thrust 2. Where the loads are limited to one direction the open U-section is ideal; this is often made into a top hat section by the addition of extra bends, and welded to panels, such as the floor, to form a closed square section. The L-section is usually limited to load-carrying

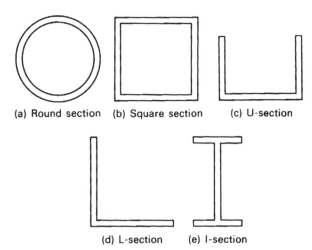

(a) Round section   (b) Square section   (c) U-section

(d) L-section   (e) I-section

FIGURE 1.8
Chassis section shapes

outriggers and trailer frames. The I-section is used for the longitudinal side rails on LGV and trailer chassis.

RACER NOTE

The tubular chassis of land speed record car Thrust 2 is made from Reynolds 531 alloy steel tube just like good quality steel bicycle frames and some motorcycle frames.

6. BODY/CHASSIS MOUNTINGS

The traditional method of bodybuilding, which is still used on some specialist cars like the Morgan, is to use sheets of steel or aluminium on a frame of hardwood such as ash. The frame is attached to the chassis. Bodies may be made from a variety of materials, including steel, aluminium, GRP, carbon fibre, and a range of other plastics materials. Modern separate bodies do not usually have a wood, or any other frame, so the body is mounted to the chassis using some form of rubber mounting. The mounting may be of the spacer and filler type, or a rubber block bonded to two plates.

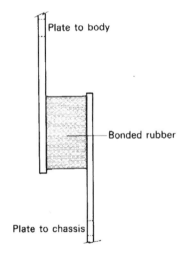

FIGURE 1.9
Body chassis mounting

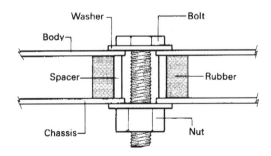

FIGURE 1.10
Body chassis mounting

7

Subframes

To reduce noise vibration and harshness (NVH) and also to speed up the vehicle assembly process, subframes are often used. The subframe may typically have the power unit and suspension attached to it; the subframe is then attached to the bodyshell using rubber mountings.

Body types

There are many different types of bodies in use, and it is important to be able to identify them quickly and accurately. This section looks at some typical styles.

Saloon — four doors, four or five seats, and a separate boot. This is sometimes called three-box construction, as from a side view you can see the passenger compartment, which looks like a box shape, with the engine compartment box to the front and the boot box at the rear. Saloons are very comfortable for carrying people separately from their luggage. Good examples are the Mercedes-Benz 180 or 200 saloons.

Hatchback — many modern vehicles are hatchbacks; these combine the passenger comfort of the saloon with the load-carrying ability of an estate car. The rear seats fold down to enable the load to be carried. They have a shorter wheelbase than the estate car, making them easy to manoeuvre in traffic. The rear door opens upwards in the form of a hatch. They usually have three doors — that is the two at the side plus the hatch door. The VW Golf was the first popular hatchback.

Estate car — also called a station wagon, a traveller and a shooting brake. It has four side doors, plus either an opening tailgate, or a pair of rear doors. It is usually slightly longer than the saloon version of the same model. Ford models such as the Mondeo and Focus are available as estate cars.

Coupé — a two-door, two-seat car with a sloping roof line. The Audi TT is a typical example.

Cabriolet — a car with four seats, two doors, and a folding hood. It is sometimes referred to as a convertible, as they convert from closed to open. The old Ford Escort Cabriolet is a good example.

Fastback — a long, low sloping back with an upward opening hatch door. Aston Martin were the first to popularize this style.

Sports car — this is a two-door, two-seat convertible. If there is a small space between the rear of the seats and the boot, then this is a roadster. This term is also misused for sports saloons and WRC/WRX cars. A sports saloon is a tuned version of a standard saloon. WRC (World Rally Car) and WRX (World Rally Cross) are cars made for special motorsport events, as are Evolution versions. MG TF and Mazda MX-5 are examples of sports cars.

Nomenclature

True sports cars are open two-seaters, like the Caterham 7; but in common language a BMW M3 may be referred to as a sports car.

FIGURE 1.11
McLaren body in white waiting to be assembled

Limousine — may have four or six doors, and six or seven seats. Extended limos are usually based on American limousines, which have their wheelbase extended even more to add another row of seats, a television lounge, or a small swimming pool.

MPV — a multi-purpose vehicle or people carrier. Door arrangements vary; they may include a sliding door and an entry rear door. Usually there are three rows of seats giving a seating capacity of seven or eight, plus some space for luggage in the same area.

Minivan — a smaller version of the MPV, usually seating a maximum of six people.

Van — a vehicle for carrying two people and a load. There are many different types of vans. Those based on the shape of a car are called **derived vans**; vans which are box shaped are called **panel vans**; the ones which have a load-carrying space above the driver are called **Luton vans**.

Off-road — for off-road use the body must be mounted high above the road to give a large amount of ground clearance; plenty of clearance between the large wheels and the wheel arches reduces the risk of fouling by mud or snow. Off-road vehicles are usually four-wheel drive; that is, 4×4 or AWD.

PCV (also called PSV) — public commercial vehicles, or as they were previously called, public service vehicles; in other words buses and coaches.

LGV (also called HGV) — large goods vehicles, or as they are also known, heavy goods vehicles. These may be of the **rigid type**, or of the **articulated type**. Articulated vehicles, called artics, have a tractive unit, usually referred to as the tractor, which houses the power unit and the driver. The tractor is attached to the trailer part by a special round shaped coupling, which is called a fifth-wheel.

7. JACKING POINTS

Jacking points are specially **reinforced** areas of the underfloor area of the body, which can be used to raise the vehicle, either with the tool kit jack, or a garage jack. This is needed to allow a wheel to be changed or other maintenance to be carried out. A typical job is the changing of brake discs. In the workshop the jacking points are often used to support the vehicle on a wheel-free device.

SAFETY NOTE

The owner's handbook, or the workshop manual, will usually show you where to find the jacking points. Under no circumstances should you ever jack a car up under the engine, the gearbox, or the fuel tank, since this is very likely to cause serious damage. You should never work underneath a car which is supported on a *hydraulic jack,* or a *tool kit jack* — always use axle stands for safety.

8. SEATS

There is a wide range of seats available. For the driver and the front seat passenger there is usually a pair of single seats. If these have a wrap-round shape they are referred to as bucket seats. The rear seat is usually a bench seat. Seats are attached to the vehicle floor using subframes. Of the two parts of the seat, the part which is normally sat on is called the cushion and the backrest, or vertical part, is called the squab.

RACER NOTE

The seat in a racing car is often moulded to match the contours of the driver's body. DIY kits are available if you wish to do this yourself on a limited budget.

9. SEAT BELTS

By law, the driver and all passengers (with a few exceptions) must use seat belts. The most popular seat belts are three-point mounting. Front seat belts usually have an inertia reel device to make them self-adjusting. For race and rally cars four- or five-point mounting seat harnesses are used.

RACER NOTE

Correctly fitting harnesses are especially needed in race and rally cars. One-off handmade harnesses are available from a small number of suppliers.

10. CORROSION PREVENTION

The load-carrying parts of car bodies are made from steel. If they are not protected from rain, or water is allowed to get onto the bare metal, then

TABLE 1.2		
Type of Prevention	**Coating Material**	**Example of Use**
Painting	Acrylic	Visible body panels
Undersealing	Rubber or wax	Underbody
Chromium plating	Chromium or nickel	Bright trim
Surface coating	Rubber or plastic	Rubbing strips
Galvanizing	Zinc	Suspension mountings

they will rust. Rust is a soft brown coating, which will eventually work its way through the metal from one side to the other. Aluminium alloy components are also affected by water; these components turn into a white powder. The term corrosion applies to both steel and aluminium alloy.

To prevent corrosion the surfaces of exposed metal components are treated with one of a number of materials.

11. INTERIOR TRIM

The interior of the vehicle is trimmed to make it comfortable to the touch for the driver and the passengers, and to reduce the level of noise. A lot of different materials are used: the top-of-the-range cars use natural materials such as **leather** for the seats, **wood** for the facia and door capping, and wool carpets. Where loads are carried, hard-wearing, resistant materials are used, such as moulded plastic and nylon carpets. Rally and race cars are not usually trimmed.

FIGURE 1.12
Classic Nomad GT racing car

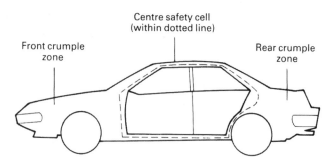

FIGURE 1.13
Body centre safety cell

12. SIPS

SIPS is the abbreviation for side impact protection system. SIPS is a set of reinforcing bars added to the side doors and vehicle side panels to protect the passengers if they were run into sideways. This is the type of accident that often happens at road junctions, when a vehicle going across a junction is hit in the side by another vehicle reaching the junction from another road. This type of accident is sometimes referred to as *T-boning*.

13. CRUMPLE ZONE

The crumple zone is the part of the front wings that are designed to compress like springs during a head-on collision. The damage to the front wings protects the passengers by softening the crash.

14. AIR BAGS

Air bags are like big floppy cloth balloons that are inflated during a crash to prevent the driver and passengers from hurting themselves on the steering wheel, windscreen and other hard parts of the car. The air bag is inflated by a small pyrotechnic device, in other words an explosive, in a time of about 2 milliseconds (0.002 seconds) — faster than you can blink.

You must follow safety procedures when working with, carrying or just storing air bags. The pyrotechnic device is triggered by a small electrical current from an inertia switch situated underneath the dashboard. Therefore you must be very careful when working on anything electrical on a vehicle with air bags.

NOTE

Air bags contain a small explosive charge — pyrotechnic — which is usually set off by an electric current. However, undue jolting, and static electricity from your overalls can trigger the charge. Remember:

- do not disconnect air bags, or anything electrical, in their vicinity unless you really know what you are doing

42 - 16   42 - 1642-50C30 FRONT FENDER

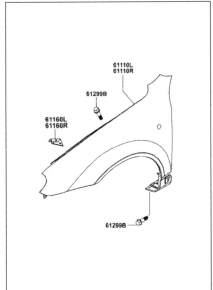

PNC	DESCRIPTION	QTY	REMARKS
61110L	FRONT FENDER LH	1	
61110R	FRONT FENDER RH	1	
61160L	BRKT ASSY FRONT FENDER LH	1	
61160R	BRKT ASSY FRONT FENDER RH	1	
61299B	BOLT, FLANGE	12	6X12

FIGURE 1.14
Front fender wing

- carry air bags carefully, and with the moving part directed away from your body
- if you remove an air bag, store it in a locked cupboard for safety.

Air bags are sometimes referred to as supplementary restraint systems (SRS); the primary restraint system is the seat belt.

13

MULTIPLE-CHOICE QUESTIONS

1. Lap and diagonal are types of:
   (a) chassis
   (b) body
   (c) seat
   (d) seat belt
2. Components known as ladder and backbone are both types of:
   (a) body
   (b) chassis
   (c) seat
   (d) paint
3. The number of seats in a sports car is usually:
   (a) one
   (b) two
   (c) three
   (d) four

4. Two types of corrosion prevention are:
   (a) painting and decorating
   (b) rubber coating and silver plating
   (c) galvanizing and painting
   (d) overcoating and undercoating

5. The vehicle layout which has a front-mounted engine driving the rear wheels is called:
   (a) FWD
   (b) RWD
   (c) conventional
   (d) normal

6. The noisiest vehicle layout is:
   (a) FWD
   (b) RWD
   (c) conventional
   (d) WRX

7. WRC is the abbreviation for:
   (a) World Rally Car
   (b) a type of truck
   (c) a type of bus
   (d) Walter Raleigh Cars

8. When carrying or storing air bags you must follow detailed safety precautions:
   (a) true
   (b) false

9. Crumple zones are found on the:
   (a) roof
   (b) front wings
   (c) sills
   (d) seat cushions

10. The system of strengthening bars used on the side panels and/or side doors of a saloon car is abbreviated to:
   (a) ABS
   (b) SRS
   (c) SIPS
   (d) FWD

(Answers on page 253.)

FURTHER STUDY

1. Draw a sketch to show how a car of your choice can be safely jacked up.
2. Look inside a car of your choice and describe how the seat adjusters work and/or how the seats move and fold.
3. Create a table to give examples of each type of vehicle layout.

Engine Technology

The engine takes in fuel and air through its inlet; it burns this mixture to create power in the form of a rotary motion at the flywheel. The amount of power that is produced by an engine depends on its size. The amount of power, and the crankshaft speed at which this power is delivered, depends on the purpose or application of the engine. For instance, the power output of a racing car engine may be the same as that of a lorry, but neither engine is suitable for the other vehicle.

There are several different types of engines in use; the most common type in the UK and the USA is the **four-stroke petrol engine**. The second most common type in the UK is the **four-stroke diesel engine**; the diesel engine is much more popular in France and other countries in continental Europe, as diesel fuel is cheaper there than in the UK. Petrol engines are also called spark ignition (**SI**) engines; diesels are called compression ignition (**CI**) engines. Both types of engines are very similar in appearance and construction; the main components are as follows.

Cylinder block (block) − this forms the main part of the engine and carries the other engine parts. The cylinder block is made from either cast iron, or, on high-performance engines, aluminium alloy. Aluminium alloy is both lighter and a better thermal conductor. The cylinder head fits onto the top of the block; the crankshaft fits into bearing housings in the lower part of the

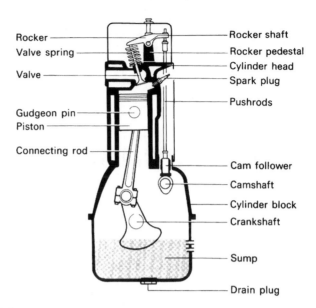

FIGURE 2.1
Engine components, for OHC engine

block. The pistons run in the cylinder bores, which are at right angles to the crankshaft. The block must be accurately machined, and very rigid, so that the components are held in exact positions relative to each other.

Pistons — these move up and down in the cylinder bores. This up and down movement is called reciprocating motion. The piston forms a gastight seal between the combustion chamber and the crankcase. The burning of the fuel and air mixture in the combustion chamber forces the piston down the cylinder to do useful work. The pistons are usually made from aluminium alloy for its light weight and excellent heat conducting

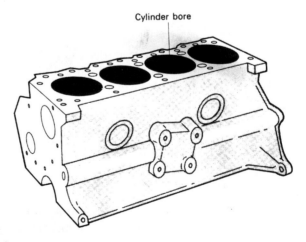

FIGURE 2.2
Cylinder block

ability. The top of the piston is called the crown; the lower part is called the skirt. The pistons must be perfectly round to give a good seal in the bore when the engine is at its normal running temperature. However, aluminium expands a lot when it is heated up. The pistons have slits in their skirts to allow for their expansion in diameter from cold to their normal operating temperature.

The pistons are fitted with piston rings to ensure a gas-tight seal between the piston and the cylinder walls. This is needed to keep the burning gases inside the combustion chamber. The piston rings are made from close grain cast iron, a metal that is very brittle. But the piston rings are slightly springy because of their shape. Usually there are three piston rings. The top two are compression rings to keep the gases in the combustion chamber. The bottom one is an oil ring; its job is to scrape the oil off the cylinder walls. The oil is returned to the sump by passing through the slots in the piston rings and running down inside the pistons. Piston rings are made from cast iron – this is very brittle so when piston rings are being fitted great care must be taken not to break them.

RACER NOTE

Race engines are developed to produce the maximum power at a specific capacity to conform with the regulations of the racing classification – this usually means high engine speed with a narrow power band.

RACER NOTE

Pistons on race engines are as light as possible to give very high-speed running – to reduce their weight the skirts are made very short and cut away to a *slipper* shape.

17

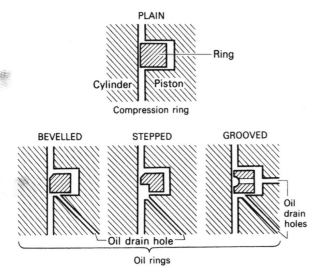

FIGURE 2.3
Types of piston rings

FAQs

What is meant by bore and stroke?

The bore is the cylindrical hole, or cylinder, in which the piston runs. The bore must be perfectly smooth, round and parallel. It is also the term used to describe the diameter of the cylinder; this is usually expressed in millimetres (mm) or inches (in). The stroke is the distance the piston travels from the bottom of the cylinder – called bottom dead centre (BDC), to the top of the cylinder – called top dead centre (TDC). The stroke may be measured in millimetres or inches. The surface inside the cylinder is called the cylinder wall.

Connecting rod (con rod) – this connects the piston to the crankshaft. The con rod has two bearings: the **little end** connects to the piston and the **big end** to the crankshaft. The con rod is made from either cast iron or forged steel. The big end bearing is a shell bearing, this allows for easy replacement and cheap manufacture.

Crankshaft – this, in conjunction with the con rod, converts the reciprocating motion of the pistons into the rotary motion which turns the flywheel. The crankshaft is located in the cylinder block by the main bearings. The big end bearings are attached to the crank pins; the crank pins are at the ends of the crank webs. The distance between the centre of the **crank pin** and the centre of the **main bearing** is called the **throw**. The throw is half of the **stroke**.

Key Points

From TDC to BDC is the stroke. For the piston to travel from TDC to BDC the crankshaft rotates 180 degrees – half a revolution. The crank pin has moved from being above the main bearing – the length of one throw – to being below the main bearing – the length of another throw. That is, two throws are equal to the length of the stroke.

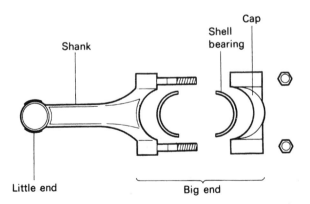

FIGURE 2.4
Connecting rod assembly

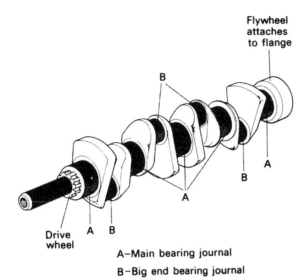

Flywheel
attaches
to flange

B

A

B

A

Drive   A   B
wheel

A–Main bearing journal

B–Big end bearing journal

FIGURE 2.5
Crankshaft for a four-cylinder engine

Cylinder head (head) — the head sits on top of the cylinder block. The head contains the combustion chambers and valves. Between the head and the block is a cylinder head gasket. The cylinder head gasket allows for the irregularities between the block and the head and keeps a **gas-tight seal** for the combustion chamber. An SI engine cylinder head locates the spark plugs; a CI engine cylinder head locates the injectors.

Valve cover and sump — the cylinder head is fitted with a valve cover (also called **rocker box**, or **cam box**). The valve cover encloses the valves and their operating mechanism, forming an oil-tight seal for the engine oil. The bottom of the block is fitted with a sump. The sump has two jobs: it is a store for the engine lubricating oil, and forms an oil-tight seal to the bottom of the engine. Both the valve cover and the sump are made from thin pressed steel.

Timing mechanism — at the front of the engine is the timing mechanism. This is either a belt or a chain, which connects the crankshaft to the camshaft. A plastic casing covers the timing mechanism. The timing end of the engine is also called the free end. Cylinder numbers always start from the free end.

Flywheel — the flywheel is attached to the crankshaft. The flywheel end of the engine is the drive end. That is, the flywheel turns the clutch and the gearbox to move the vehicle.

19

Key Points

As a race mechanic, it is good to understand the different metals which are used in engines. Their properties must be considered when you are handling the part, and particularly when tightening up nuts and bolts.

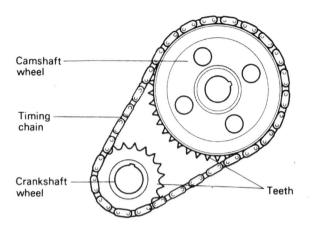

FIGURE 2.6
Camshaft timing gear ratio

- Cast iron is used for many cylinder blocks and heads; it is very heavy and brittle so will break if dropped.

- Aluminium is light in weight and expands a great deal when heated up. It is also soft, so it is easily scratched. You must be careful not to over-tighten spark plugs in aluminium cylinder heads or you will damage the threads.

- Pressed steel is used for sumps and valve covers; this is easily bent. A bent sump may leak around the joints.

- Hardened steel is used for the crankshaft; this is both heavy and expensive.

1. FOUR-STROKE PETROL ENGINE

The four-stroke petrol engine works on a cycle of four up and down movements of the piston. These up and down movements are called strokes. The piston moves down from top dead centre (TDC) to bottom dead centre (BDC), then up to TDC again. Each stroke corresponds to half of a turn of the crankshaft; therefore the complete cycle of four strokes takes two revolutions of the crankshaft.

The petrol and air mixture is burnt in the combustion chamber during one of the strokes. The heat from the burning fuel causes a pressure increase in the combustion chamber. This pressure forces the piston down the bore to do useful work. The mixture is ignited by the spark plug, hence the term spark ignition (SI).

The cylinder head is fitted with inlet valves; these open and close to control the flow of the petrol and air mixture from the inlet manifold into the combustion chamber. The cylinder head is also fitted with exhaust valves to control the flow of the spent exhaust from the combustion chamber into the exhaust manifold and exhaust system. The passage in the cylinder head, which connects the manifold to the combustion chamber, is called the port. There are inlet ports and exhaust ports. The valves are situated where the ports connect into the combustion chamber. The valves are operated by the camshaft, which is discussed later in this chapter.

Induction stroke

Key Points

- *Inlet valve open*
- Exhaust valve closed
- *Piston travelling from TDC to BDC*

The piston travels down the cylinder bore from TDC, drawing in the mixture of petrol and air from the inlet manifold. This is like a syringe drawing up a liquid. The downward movement of the piston has caused a depression above the piston. This depression, or partial vacuum, is satisfied by the air coming into the inlet manifold through the air filter. The air mixes with the petrol that is supplied from either the injectors or a carburettor.

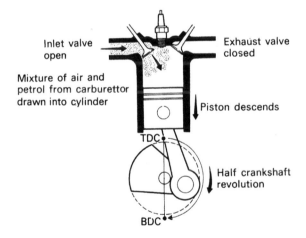

Inlet valve
open

Exhaust valve
closed

Mixture of air and
petrol from carburettor
drawn into cylinder

Piston descends

TDC

Half crankshaft
revolution

BDC

FIGURE 2.7
4-stroke engine — induction stroke

Compression stroke

Key Points

- *Inlet valve closed*
- Exhaust valve closed
- *Piston travelling from BDC to TDC*

21

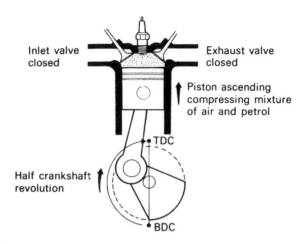

Inlet valve
closed

Exhaust valve
closed

Piston ascending
compressing mixture
of air and petrol

TDC

Half crankshaft
revolution

BDC

FIGURE 2.8
4-stroke engine — compression stroke

When the piston reaches BDC it starts to return up the bore. At about BDC the inlet valve is closed by the camshaft — the exhaust valve was already closed. The mixture of petrol and air, which was drawn in on the induction stroke, is now compressed into the combustion chamber. This increases the pressure of the mixture to about 1250 kPa (180 psi). The actual pressure depends on the compression ration of the engine — on race engines it is typically between 10:1 and 16:1.

Nomenclature
The mathematical sign **:** signifies a ratio. For example 9:1 is a ratio, it is said 'nine to one'.

Power stroke

Key Points

- *Inlet valve closed*

- Exhaust valve closed

- *Piston travelling from TDC to BDC*

As the piston reaches TDC on the compression stroke, the spark occurs at the spark plug. This spark, which is more than 10 kV (10 000 volts), ignites the petrol/air mixture. The mixture burns at a temperature of over 2000 degrees Celsius and raises the pressure in the combustion chamber to over 5000 kPa (750 psi). The pressure of the burning petrol/air mixture now starts to force the piston back down the cylinder bore to do useful work. The piston rings seal the pressure of the burning mixture into the combustion chamber, so that it exerts a force on the piston, the gudgeon pin and then the con rod, which converts this downward motion into rotary motion at the crankshaft.

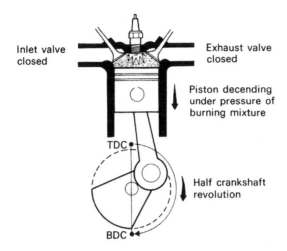

Inlet valve
closed

Exhaust valve
closed

Piston decending
under pressure of
burning mixture

TDC

Half crankshaft
revolution

BDC

FIGURE 2.9
4-stroke engine — power stroke

FAQs

How much force does the burning mixture actually exert on the con rod or crankshaft and why do they not bend?

The amount of force depends on the size of the engine; but as a rough guide, imagine an elephant sat on the top of the piston every time it goes down. The components will not bend as long as the engine is rotating and the force is being passed onto the transmission to move the vehicle.

Exhaust stroke

Key Points

- *Inlet valve closed*

- Exhaust valve open

- *Piston travelling from BDC to TDC*

At the end of the power stroke the exhaust valve opens. When the piston starts to ascend on the exhaust stroke, this is the last stroke in the cycle; the burnt mixture is forced out into the exhaust. The mixture of petrol and air has been burnt to change its composition. Its energy has been spent. The temperature of the exhaust gas is about 800 degrees Celsius. The petrol/air mixture has been burnt to become carbon monoxide (CO), carbon dioxide (CO_2), water (H_2O), nitrogen (N) and free carbon (C). The exhaust gas is passed through the exhaust system to the catalytic converter to be cleaned and made non-toxic.

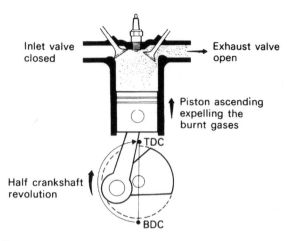

Inlet valve closed

Exhaust valve open

Piston ascending expelling the burnt gases

TDC

Half crankshaft revolution

BDC

FIGURE 2.10
4-stroke engine — exhaust stroke

Flywheel inertia

There is only one firing stroke for each cycle. The flywheel keeps the engine turning between firing strokes. Single cylinder engines need a bigger flywheel in proportion to their size than those with more cylinders. The flywheel on a large V8 engine is smaller than one on a four-cylinder engine. The flywheel's desire to keep rotating is called inertia; it is inertia of motion.

2. FOUR-STROKE DIESEL ENGINE

The operation of the four-stroke diesel engine is very similar to the four-stroke petrol engine. The diesel engine draws in air only, and then

FIGURE 2.11
New Lola T70 racing car

compresses this to a very high pressure and temperature, which causes the fuel to combust and burn. The fuel is injected directly into the engine at a very high pressure. Vehicles with diesel engines are very economical and they produce lots of pulling power at low engine speeds.

RACER NOTE

It should be remembered that Audi won the Le Mans 24 hour race with a diesel engined car.

Induction stroke

Key Points

- *Inlet valve open*
- Exhaust valve closed
- *Piston travelling from TDC to BDC*

The piston travels down the cylinder bore from TDC drawing in **air only** from the inlet manifold. This is like a syringe drawing up a liquid. The downward movement of the piston has caused a depression above the piston, this depression, or partial vacuum, is satisfied by the air coming into the inlet manifold through the air filter.

Compression stroke

Key Points

- *Inlet valve closed*
- Exhaust valve closed
- *Piston travelling from BDC to TDC*

When the piston reaches BDC it starts to return up the bore. At about BDC the inlet valve is closed by the camshaft — the exhaust valve was already closed. The air, which was drawn in on the induction stroke, is now compressed into the combustion chamber. This increases the pressure of the air to about

25

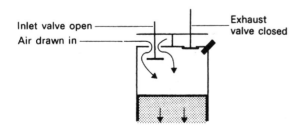

Inlet valve open
Air drawn in
Exhaust valve closed

FIGURE 2.12
Diesel — induction stroke

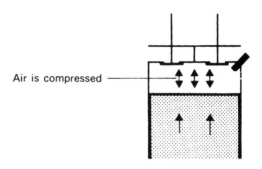

Air is compressed

FIGURE 2.13
Diesel — compression stroke

2000 kPa (300 psi). The actual pressure depends on the compression ratio of the engine. The compression ratio of a CI engine is much higher than that of an SI engine; it may exceed 22:1. This increase in pressure causes an increase in temperature. The compression temperature inside the combustion chamber just before TDC reaches about 600 degrees Celsius.

Power stroke

> ### Key Points
>
> - *Inlet valve closed*
> - Exhaust valve closed
> - *Piston travelling from TDC to BDC*

As the piston reaches TDC on the compression stroke, the diesel fuel is injected into the combustion chamber. As the air temperature in the combustion chamber (the compression temperature) is about 600 degrees Celsius, the fuel starts to burn — hence the name compression ignition (CI). The diesel fuel and air burn at a similar temperature to the mixture in the petrol engine, i.e. over 2000 degrees Celsius. This raises the pressure in the combustion chamber. The pressure of the burning diesel/air mixture now starts to force the piston back down the cylinder bore to do useful work. The piston rings seal the pressure of the burning mixture into the combustion chamber so that the pressure exerts a force on the piston, the gudgeon pin and then the con rod, which converts this downward motion into rotary motion at the crankshaft.

Exhaust stroke

> ### Key Points
>
> - *Inlet valve closed*
> - Exhaust valve open
> - *Piston travelling from BDC to TDC*

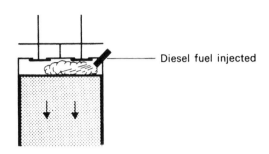

FIGURE 2.14
Diesel — power stroke

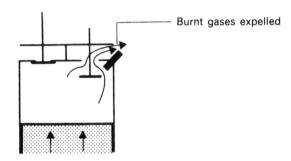

FIGURE 2.15
Diesel — exhaust stroke

At the end of the power stroke the exhaust valve opens. When the piston starts to ascend on the exhaust stroke, this is the last stroke in the cycle; the burnt mixture is forced out into the exhaust. The mixture has been burnt to change its composition. Its energy has been spent. The temperature of the exhaust gas is about 800 degrees Celsius. The diesel/air mixture has been burnt to become the hot exhaust gas. Diesel exhaust gas contains more CO_2 than CO; it also contains more carbon particulate.

Flywheel inertia

Because of their high compression ratios, and heavy moving parts, diesel engines usually have large flywheels.

FAQs

How can I tell the difference between a petrol engine and a diesel engine?
Both engines look similar to each other, but the following points will help you to spot which one is which.

- Diesel engines do not have spark plugs, they have injectors.
- Diesel engine components are much heavier than petrol ones to cope with the higher compression ratio.
- Diesel engines run slightly slower than petrol engines and make a louder noise.
- The fuel for the two engines smell differently.

Valve operation

The inlet and the exhaust valves each open once every two revolutions of the crankshaft. The mechanism for opening the valves is a camshaft, which either acts directly on the valves or operates them through a pushrod and a rocker shaft assembly.

If the camshaft is situated above the valves the engine is referred to as overhead cam (OHC). Other types are overhead valve (OHV) and side valve. A rubber-toothed belt or a chain may drive the camshaft. As the camshaft must rotate at half the speed of the crankshaft, to open the valves once for every two revolutions of the crankshaft, the drive is through a two to one (2:1) gear ratio. The cam gear wheel has twice as many teeth as the crankshaft gear wheel.

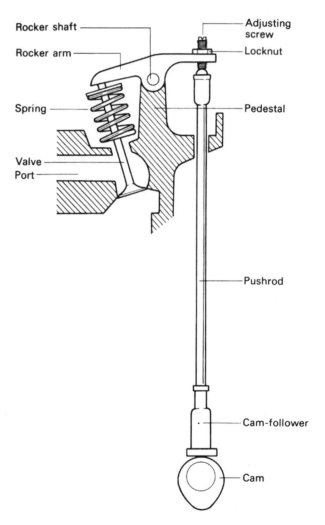

FIGURE 2.16
OHV arrangement

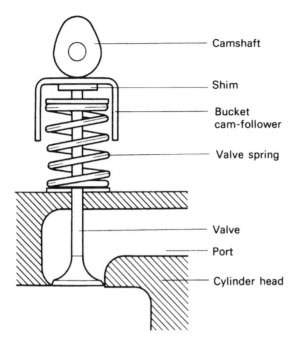

FIGURE 2.17
OHC arrangement

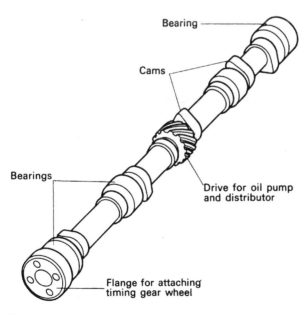

FIGURE 2.18
Camshaft

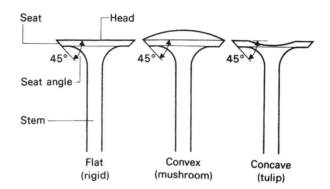

The three types of valves most commonly used

FIGURE 2.19
Valve shapes

Nomenclature

You may be tempted to say sprocket when you mean gear wheel. On a chain drive the term is sprocket or gear wheel; with toothed belts it is definitely gear wheel. Gearboxes only have gear wheels. Bicycles have chain rings, chains and sprockets.

The actual point at which the valves open and close depends on the engine design. The vehicle's workshop manual will give the valve timing figures; these are usually expressed in degrees of the crankshaft relative to TDC and BDC.

30

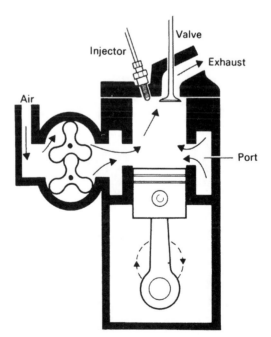

FIGURE 2.20
2-stroke diesel BDC

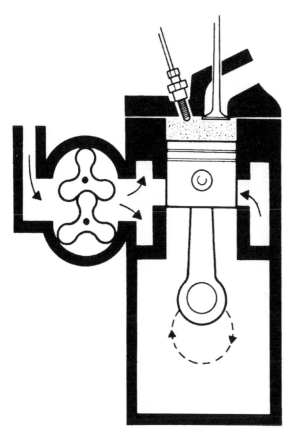

FIGURE 2.21
2-stroke diesel TDC

It is essential that the valves close firmly against their seats to give a good, gas-tight seal, and allow heat to conduct from them to the cylinder head so that they may cool down. The valves are held closed by springs. To ensure that the valves close firmly, even when the components have expanded because of the heat, the valve mechanism is given a small amount of clearance. The valve clearance is measured with a feeler gauge; a typical figure is 0.15 mm (0.006 in). If the valve clearance is too great then a light metallic rattling noise will be heard.

3. TWO-STROKE PETROL ENGINE

The two-stroke petrol engine is used mainly in small motorcycles, although they have been used in some cars. It operates on one up-stroke and one down-stroke of the piston; that is one revolution of the crankshaft. The most common type is the Clerk cycle engine, which has no valves, just three ports. The three ports are the inlet port, the transfer port and the exhaust port. The flow of the gas through these ports is controlled by the position of the piston. When the piston is at TDC both the transfer and the

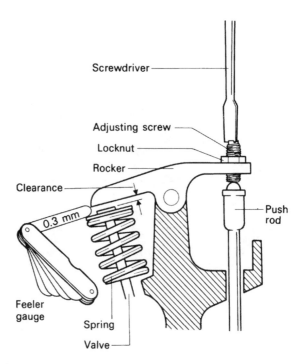

FIGURE 2.22
Valve clearance adjustment

exhaust ports are closed. When the piston is at BDC the piston skirt closes the inlet port.

The piston travels up the bore. As it reaches TDC it closes both the transfer and the exhaust ports. At the same time, the piston is compressing the charge of petrol and air above it into the combustion chamber. At about TDC the spark plug ignites the petrol/air mixture. The burning of the petrol/air mixture increases the temperature and the pressure so that the burning gas pushes the piston down the bore. The downward force of the piston is passed through the gudgeon pin to the con rod and crankshaft to drive the vehicle.

Whilst the piston is ascending, its skirt uncovers the inlet port. The upward motion of the piston causes a vacuum in the crankcase, which is satisfied by the petrol/air mixture from the carburettor entering through the now open inlet port.

When the piston is travelling downwards — being forced down by the burning mixture — the piston crown first uncovers the exhaust port. This allows the spent gas to escape into the exhaust system. The skirt of the piston covers the inlet port at the same time as the piston crown uncovers the transfer port at the top of the cylinder. The underside of the piston therefore acts like a pump plunger, forcing the fresh charge of petrol/air that is in the crankcase up and through the transfer port into the cylinder so that another cycle is started.

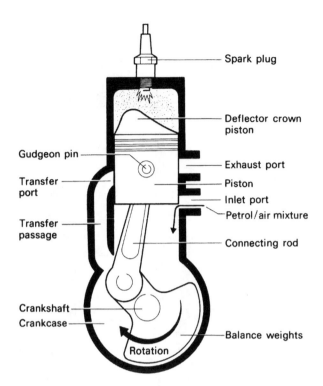

Spark plug

Deflector crown
piston

Gudgeon pin

Transfer
port

Transfer
passage

Exhaust port

Piston

Inlet port

Petrol/air mixture

Connecting rod

Crankshaft

Crankcase

Balance weights

Rotation

FIGURE 2.23
2-stroke engine — piston ascending

The two-stroke petrol engine is much lighter and simpler than the four-stroke petrol engine and has fewer moving parts. It has neither valves nor a valve operating mechanism. Two-stroke petrol engines usually run at high speeds. The problem is that oil must be mixed with the petrol for lubrication, and this causes the exhaust to smoke.

RACER NOTE

Two-stroke engines were used in the early karts and there are still racing classes for these, however, because of emission regulations no new ones are being made. Old racing two-stroke motorcycles were loved by many for the smell of the burnt oil — even though it could be a health hazard. The oil was vegetable-based, and called *Castol R* — it is currently available for historic applications.

4. FIRING ORDER

By increasing the number of cylinders the engine becomes more compact and smoother running for its size. Smoothness of running is further improved by setting the sequence in which the cylinders fire; this is called the firing order. The normal firing orders for four-cylinder engines are 1-3-4-2 and 1-2-4-3.

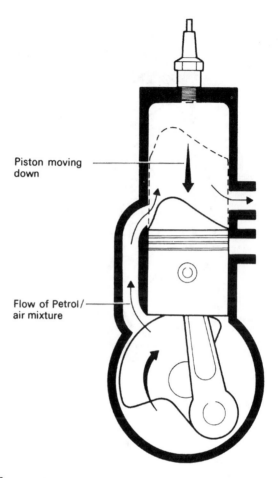

FIGURE 2.24
2-stroke engine — piston descending

5. ENGINE CAPACITY

To find the capacity of an engine first you need to find the size of each cylinder; this is called the swept volume. This is the volume that is displaced when the piston goes from BDC to TDC. The swept volume of each cylinder is obtained by multiplying the cross sectional area by the stroke.

$$\text{Swept Volume} = \Pi D^2 L/4$$

$\Pi = 3.142$

D = Diameter of Bore

L = Length of Stroke

All divided by 4

Key Points

These sums are very easy once you have done one. Get your tutor to show you how to do them using either a calculator or a spreadsheet. It is really

FIGURE 2.25
Maserati 250F

useful when you come to swap engines and identify unknown cylinder blocks.

The engine capacity is the product of the swept volume and the number of cylinders.

Capacity = Swept Volume × Number of Cylinders

You will find the following abbreviations useful:
V_S = Swept Volume
N = Number of Cylinders

The engine capacity is usually measured in cubic centimetres (cc).

Nomenclature
There are 1000 cc in 1 litre. A 1000 cc car is therefore referred to as one litre. American cars are sized in cubic inches (cu in). 1 litre is equal to 62.5 cu in.

6. COMPRESSION RATIO

The compression ratio is the relationship between the volume of gas above the piston at BDC compared to that at TDC. You need to know the swept volume (V_s) and the volume of the combustion chamber that is referred to in this case as the clearance volume (Vc).

$$\text{Compression Ratio} = V_s + V_c/V_c$$

Petrol engines have compression ratios of about 9:1; diesel engines are usually over 14:1.

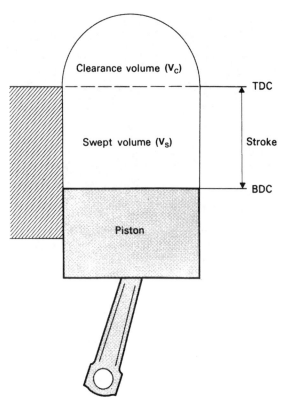

FIGURE 2.26
Compression ratio

36

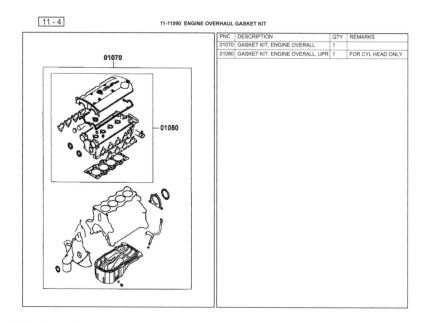

FIGURE 2.27
Engine overhaul gasket set

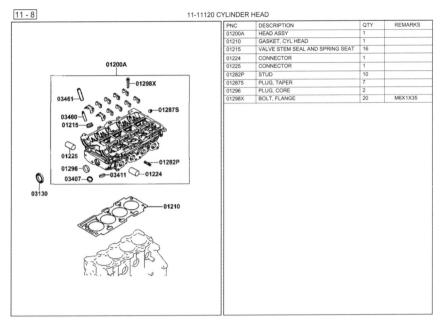

PNC	DESCRIPTION	QTY	REMARKS
01200A	HEAD ASSY	1	
01210	GASKET, CYL HEAD	1	
01215	VALVE STEM SEAL AND SPRING SEAT	16	
01224	CONNECTOR	1	
01225	CONNECTOR	1	
01282P	STUD	10	
012875	PLUG, TAPER	7	
01296	PLUG, CORE	2	
01298X	BOLT, FLANGE	20	M6X1X35

11-11120 CYLINDER HEAD

11 - 8

FIGURE 2.28
Cylinder head

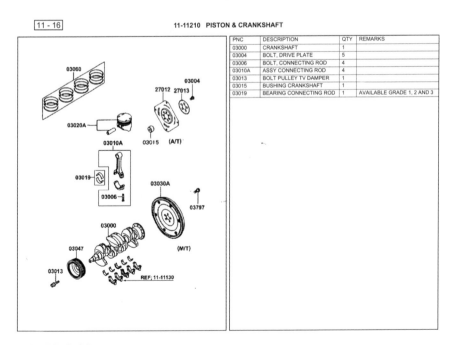

PNC	DESCRIPTION	QTY	REMARKS
03000	CRANKSHAFT	1	
03004	BOLT, DRIVE PLATE	5	
03006	BOLT, CONNECTING ROD	4	
03010A	ASSY CONNECTING ROD	4	
03013	BOLT PULLEY TV DAMPER	1	
03015	BUSHING CRANKSHAFT	1	
03019	BEARING CONNECTING ROD	1	AVAILABLE GRADE 1, 2 AND 3

11-11210 PISTON & CRANKSHAFT

11 - 16

FIGURE 2.29
Piston and crankshaft

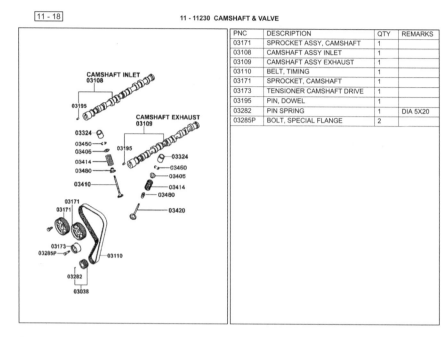

11 - 18				11 - 11230 **CAMSHAFT & VALVE**

PNC	DESCRIPTION	QTY	REMARKS
03171	SPROCKET ASSY, CAMSHAFT	1	
03108	CAMSHAFT ASSY INLET	1	
03109	CAMSHAFT ASSY EXHAUST	1	
03110	BELT, TIMING	1	
03171	SPROCKET, CAMSHAFT	1	
03173	TENSIONER CAMSHAFT DRIVE	1	
03195	PIN, DOWEL	1	
03282	PIN SPRING	1	DIA 5X20
03285P	BOLT, SPECIAL FLANGE	2	

FIGURE 2.30
Camshaft and valve

MULTIPLE-CHOICE QUESTIONS

1. The first and second strokes of an SI engine are:
- **(a)** power and exhaust
- **(b)** power and compression
- **(c)** induction and compression
- **(d)** exhaust and compression

2. The flywheel is said to store:
- **(a)** oil
- **(b)** fuel
- **(c)** power
- **(d)** inertia

3. On the induction stroke the SI engine draws in a mixture of petrol and air. The CI engine draws in:
- **(a)** air only
- **(b)** fuel only
- **(c)** petrol only
- **(d)** diesel fuel and air

4. The gear ratio between the crankshaft and the camshaft is:
- **(a)** 1:1
- **(b)** 2:1
- **(c)** 3:1
- **(d)** 4:1

5. The temperature inside a combustion chamber at the start of the power stroke is:
   (a) 85 degrees Celsius
   (b) 600 degrees Celsius
   (c) 800 degrees Celsius
   (d) 2000 degrees Celsius

6. The name of the component which connects the piston to the connecting rod is the:
   (a) cotter pin
   (b) split pin
   (c) clevis pin
   (d) gudgeon pin

7. A typical firing order for a four-cylinder engine is:
   (a) 1-3-4-2
   (b) 1-2-3-4
   (c) 1-3-4-2
   (d) 1-4-3-2

8. The part of the cylinder head which is between the manifold and the valve is the:
   (a) widget
   (b) neck
   (c) venturi
   (d) port

9. The spark plug is screwed into the:
   (a) sump
   (b) cylinder head
   (c) piston
   (d) crankshaft web

10. The cylinder head gasket allows for roughness in the surface of the:
   (a) cylinder head
   (b) sump
   (c) manifold
   (d) crankshaft

(Answers on page 253.)

FURTHER STUDY

1. Why are some vehicles powered by petrol engines and others by diesel engines?
2. Investigate a number of engines and identify which are OHC and which are OHV.
3. Calculate the swept volume and the capacity of an engine of your choice. You may find the figures in a workshop manual.

CHAPTER 3

Fuel System

KEY POINTS

- Both petrol and diesel fuel must be handled with care; they are highly flammable and can be a hazard to your health.
- Petrol engines mainly use fuel injection, but some older vehicles have carburettors.
- Diesel fuel is injected into the combustion chamber at a pressure of about 170 bar (2500 psi).
- Air filters and fuel filters need changing regularly.
- Diesel fuel system components must be handled with care to prevent personal injury.
- The fuel system must be set up to supply the correct amount of fuel at the correct time.
- Correct servicing is needed to maintain both fuel economy and reduce environmental pollution.

The fuel system supplies the motor vehicle with the necessary amount of fuel for it to be able to do its work efficiently. The engine must receive the correct amount of fuel at the right time or else it will not run properly, if at all. Fuel is the food of the engine. Most vehicles run on either petrol or diesel fuel which is bought in liquid form. Both petrol and diesel fuel are hydrocarbons, as they are made up of hydrogen and carbon atoms; but the similarity ends there. You cannot run a petrol engine on diesel fuel, or vice versa. Indeed, just putting the wrong fuel in the tank can cause a lot of expensive damage.

It is important to remember that burning fuel creates a hot exhaust gas, which, if the fuel system is not set up correctly, can cause illegal pollution.

Working on a fuel system, either petrol or diesel, presents the mechanic with a number of dangers and hazards. Remember to follow the correct safety procedures and you will be quite safe. Let's look at the petrol system first.

1. PETROL SUPPLY SYSTEM

SAFETY NOTES

Before working on a petrol supply system you should be aware of the following hazards:

- Petrol is highly flammable; do not smoke or have any naked lights near petrol.
- Always wear protective gloves and avoid direct contact with petrol as it dries the skin and prolonged exposure can cause skin disease.
- If your overalls are doused in petrol at all change them; there is an extreme fire risk even after the petrol has dried.
- If draining a fuel tank, a sealed and electrically earthed draining appliance must be used.
- Petrol must not be stored in the garage.

Basic carburettor petrol supply system

This section covers the basic carburettor system found on older vehicles, kit cars and special vehicles.

The basic system, uses the carburettor to mix the petrol and air in the correct proportions. The main components are the petrol tank, the petrol pump, the carburettor and the air filter. Let's briefly look at the components in turn.

42

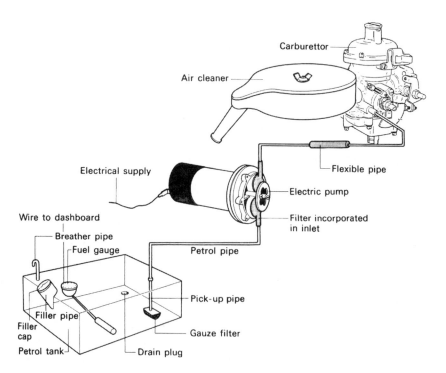

FIGURE 3.1

Fuel system layout, for injection system

Petrol tank — usually this is situated underneath the car and behind the driver and the passengers to reduce the risk of its contents spilling over them in the event of an accident. Petrol tanks may be made either from pressed low-carbon steel or moulded plastic. Plastic petrol tanks are lighter and do not rust. The capacity of the petrol tank depends on the expected use of the vehicle and the average fuel consumption. For example, if the vehicle has a fuel consumption of 10 litres per 100 kilometres and its average daily usage is expected to be 500 kilometres, then:

$$\text{Fuel tank capacity} = 500/100{*}10$$
$$= 50 \text{ litres}$$

The petrol tank is fitted with a pick-up pipe that is just above the lowest point of the tank so that it does not pick up any sediment. The filler cap must seal against the tank. The breather is from the highest part of the tank, so not to allow fuel to leak out; the open end of the breather is at a low point at the rear of the vehicle so that fumes do not enter the passenger compartment. The petrol tank is also fitted with a floating sensor to measure the amount of petrol in the tank. The sensor sends an electrical signal to the dashboard-mounted gauge. The gauge may record the level of the petrol in relative terms, such as full, half full and empty, or in actual litres.

Nomenclature

Fuel consumption for cars is given in SI units as litres per hundred kilometres (litres/100 km). In imperial units it is in miles per gallon (mpg). For conversion purposes 9 litres/100 km approximately equals 30 mpg.

Fuel (petrol) pump — this may be operated mechanically from the camshaft, or electrically through the ignition switch. The mechanical petrol pump is self-regulating. That is, it uses the force of the spring to increase the petrol's pressure. If the carburettor does not need petrol, the mechanism freewheels, the spring holding a constant pressure. The diaphragm-type electrical pump operates in the same way.

The petrol pump draws up petrol from the tank and sends it, under pressure, to the carburettor. The fuel lines, or pipes, may be made from steel or plastic. The pipes are usually a push fit onto the petrol tank pickup, the pump and the carburettor. Where the pipes run along the side of the chassis, or across parts of the body, they are clipped to hold them in place and stop chafing. If the petrol pipe is chafed through then the vehicle will not run, or worse still, the fuel that leaks out could cause a fire.

FAQs

How can I tell if the petrol pump is faulty?

The pump supplies petrol to the carburettor at a pressure of about 30 kPa (5 psi), if the pump is suspected of being faulty the pressure can be measured with a gauge.

43

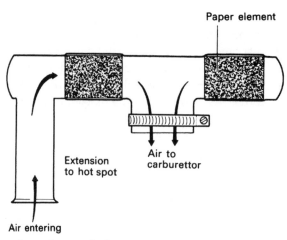

Paper element air cleaner

FIGURE 3.2
Air filter

Carburettor − this mixes the petrol and the air in the correct proportions for it to be burnt inside the combustion chamber.

The simple carburettor

The basic principles of carburation are embodied in the simple carburettor. The simple carburettor was used on the very first cars and motorcycles and can now be found on lawnmowers and other single-speed garden equipment. The petrol is in the **float chamber**, which is connected to the **discharge jet** through a tube that contains the main metering jet. When the engine is spun over with the throttle valve and the **choke butterfly** open, air is drawn into the **venturi**. As the air passes through the narrow section of the venturi, its speed increases. Increasing the air speed causes the air pressure to drop, so that the air pressure in the venturi is below normal atmospheric pressure. The petrol in the float chamber is at atmospheric pressure; it now flows through the main metering jet and out of the discharge nozzle to mix with the air. As the petrol leaves the discharge nozzle it breaks into very small droplets − this is referred to as **atomization**.

Nomenclature
Venturi is the name given to the narrowing of an air passage. In a carburettor this is sometimes referred to as the choke tube, or simply choke. High performance Weber carburettors usually have two chokes; American Holley carburettors have four chokes.

The **throttle flap** controls the flow of petrol and air through the carburettor; on a car this is operated by the throttle pedal. The **choke flap** controls the mixture strength by limiting the airflow into the carburettor. The choke is closed for cold starting to give a richer mixture.

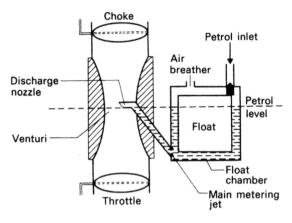

FIGURE 3.3
Simple carburettor

The flow of petrol into the carburettor float chamber is controlled by the **needle valve** and float. As the **float chamber** fills up with petrol, the float rises. The rising float pushes the needle valve up against its seat and stops the flow of petrol.

Air to petrol ratio

For the complete combustion of petrol and air, 14 parts of air are needed to burn one part of petrol. However, when an engine is running under different conditions the mixture strength needs to be altered. Table 3.1 shows typical mixture strengths for different operating conditions.

Adjusting the carburettor – you can usually adjust both the **mixture strength** and the **idling speed** of the carburettor. The workshop manual will give information on both of these settings.

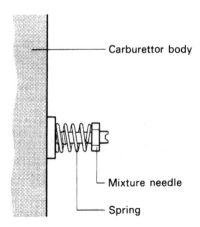

FIGURE 3.4
Mixture adjustment

45

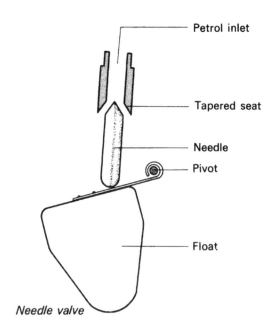

FIGURE 3.5
Float height

FAQs

How are carburettor adjustments checked?
The idling speed on most carburettor engines is between 600 and 900 rpm. Generally if it sounds nice it is correct. Because of the exhaust emission regulations you must set the mixture using an exhaust gas analyser — cheap toolbox ones are available for the home mechanic or the school on a tight budget.

Air filter

The air filter assembly has three functions, namely:

1. To filter and clean the incoming air so preventing the entry of dust, grit and other foreign bodies into the engine, which could seriously damage it.
2. To silence the air movement, so making the engine less noisy.
3. To act as a flame trap, so preventing a serious under-bonnet fire should the engine backfire.

TABLE 3.1	
Operating Condition	**Air:Petrol Ratio**
Cold starting	9:1
Slow running	13:1
Accelerating	11:1
Cruising	19:1

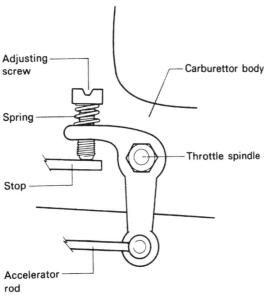

Adjusting screw

Carburettor body

Spring

Throttle spindle

Stop

Accelerator rod

Adjusting idling speed. To speed up the engine, turn the screw clockwise

FIGURE 3.6
Throttle adjustment

Most vehicles are fitted with paper element air filters. These must be replaced at set mileage intervals — usually about 20 000 miles. Air filters must be replaced regularly even if they look clean; their micropore surface may still be blocked.

RACER NOTE

Replacing the standard paper air filter with a mesh type, such as the K&N version, may increase the engine power output.

Petrol injection

This section covers the petrol injection system used on most new vehicles — variations are used for specialist applications.

All new cars are fitted with petrol injection (PI). Petrol injection cars have a similar petrol tank, petrol pipes, and air filter. The carburettor is replaced by injectors and a high-pressure pump. PI systems are usually more efficient than carburettors in that they give more power, better fuel consumption and more controllable emissions; they also need less maintenance. Let's have a look at the main components.

FUEL (PETROL) PUMP

The petrol pump on a PI engine is likely to be electrically operated and submerged in the petrol tank. It is submerged to keep it cool and prevent the entry of air bubbles into the petrol pipes. The PI engine petrol pump raises the petrol pressure to about 1650 kPa (110 psi) and sends it to the injectors via a filter. The petrol pipes on a PI system are much heavier than those on a carburettor engine car, the pipe couplings are by screw threaded ends. The petrol pump is controlled electronically by the Electronic Control Unit (ECU).

ELECTRONIC CONTROL UNIT (ECU)

The ECU is a sealed box containing a number of microprocessor integrated circuits — usually referred to as microchips — similar to those in a computer. The ECU controls both the idle speed and the mixture strength, so there is no need to adjust these settings. The ECU for the petrol injection equipment also controls the ignition system.

INJECTORS

There may be one centrally mounted injector, called single-point injection, or one for each cylinder. The petrol is supplied under pressure to the injectors from the petrol pump. The ECU electronically controls the amount of fuel to be delivered by the injectors.

Injectors open and close about 50 times each second when the engine is running at full speed, on each occasion delivering petrol. Injectors can become blocked or worn. To maintain injectors in good working order they should be removed and checked at regular service intervals. Their spray pattern is observed and they may be cleaned or replaced as needed. Cleaning may be by ultrasonic vibrations or the use of a chemical cleaning agent. Injector cleaning chemicals are available to be added to the petrol in the tank.

Nomenclature

In this book we have used the term petrol injection and the abbreviation PI because it is a general term. You will come across Electronic Fuel Injection (EFI); just Injection (I, or i); Gasoline Injection (GI) and a range of equivalent terms in different languages.

AIR FILTER

On a PI engine this is similar to the one used on carburettor engine cars.

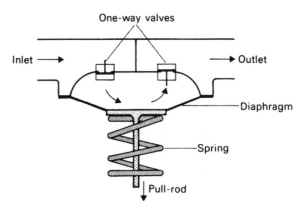

FIGURE 3.7
Fuel pump

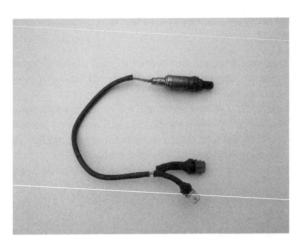

FIGURE 3.8
Lamda sensor

AIR FLOW CONTROL

To control the flow of air into the engine, and hence the engine speed,
a throttle body similar to that in a carburettor is used. This is connected by
the accelerator cable to the accelerator (or throttle) pedal. The throttle
butterfly is connected electrically to the ECU.

Fuel cell

On race cars it is essential to store the fuel safely. The petrol used in race cars
tends to be very volatile as it easily ignites; it also burns to release more
energy than that used in road cars. In other words it is very hazardous and
needs stringent controls to keep it safe. To this end a fuel cell is used instead
of a normal tank. The fuel cell contains a foam material which in effect
absorbs the petrol and prevents it from spilling or leaking if the tank is
inverted or damaged. The fuel cell is filled in the same way as a standard tank,
through a large filler cap — usually these are connected to the filler hose with
a sealed push-and-twist type of coupling.

RACER NOTE

A number of different fuels are used for race cars — for certain series of
racing a control fuel is used — one which is available only from one source.

2. DIESEL FUEL SUPPLY

The main difference between the petrol supply system and the diesel fuel
supply system is that the petrol system takes in a mixture of petrol and air
together; whereas the diesel system draws only air into the engine and the
diesel fuel is injected separately into the combustion chamber.

49

SAFETY NOTES

- Diesel fuel may look clean, but it contains materials which can cause dermatitis or other skin problems — always use barrier cream and rubber gloves when working on diesel engine components.
- The internal parts of the injector pump and the injectors will be damaged if touched directly by your hands — always handle the parts while they are submerged under the special oil which is available for this — do not substitute other oils or diesel fuel as they may damage both your skin and the parts.
- If you spill diesel fuel on your skin, wash it off immediately and reapply your barrier cream.
- You must change your overalls regularly, and immediately if you spill diesel fuel on them.
- Do not keep rags which have diesel fuel on them in your pocket: prolonged exposure can lead to skin disease.

Diesel fuel tank

On a diesel engine car the fuel tank will be very much like the petrol tank on the petrol version of the same model. The filler neck is likely to be marked *Use Diesel Fuel Only,* or with similar wording. Diesel fuel is sometimes referred to as DERV, or gas oil.

Diesel trucks have fuel tanks of at least 250 litres (50 gallons). This is because they have high fuel consumption and cover large distances. For instance if a truck's fuel consumption was 10 mpg, a 50 gallon tank would allow just less than 500 miles between refills — you cannot completely drain the tank.

NOTE

You should never allow a diesel vehicle to run out of fuel. If you do this the system will need venting (bleeding) and the pumping components may be damaged.

Air filter

The air filter on a diesel engine is similar to that on a petrol engine. However, they are usually larger to allow in a greater amount of air. It is important that the air filter is replaced at the recommended service intervals.

Lift pump

The diesel engine has a lift pump to draw fuel from the tank to the injector pump via the fuel filter. The lift pump operates at about 30 kPa (5 psi), which is similar to the petrol pump on a carburettor engine. The diesel fuel lift pump is fitted with an external handle so that it can be operated manually when the engine is stationary. This is used for priming the fuel system before starting the engine. See the section on venting for more information.

Fuel filter

Because it is essential that the fuel is very clean before it goes to the injector pump and the injectors, it is passed through a very comprehensive filter. The filter uses a very fine paper element. The filter housing is also fitted with a mechanical trap to separate the fuel and any water, and an area for sediment to settle into.

Injector pump

There are two main types of injector pumps:

- Inline injector pump
- Distributor, or rotary, injector pump — often called a DPA pump, an abbreviation of distributor pump assembly.

The inline pump is used on larger diesel engines — those used in trucks and buses. The DPA pump is used almost universally on car and light goods vehicles.

The injector pump sends fuel to the injectors at a pressure of about 170 bar (2500 psi). Because of the extremely high pressure, special, thick-walled pipes are needed; these are referred to as injector pipes. The injector pipes are connected to the pump at one end and the injectors at the other end, using special couplings called glands. It is usual to use special slotted hexagonal spanners to tighten up these coupling glands. These spanners will not slip on the gland nuts.

Injectors

The injector is fitted into the cylinder head. It projects into the combustion chamber so that the fuel can be injected into the hot compressed air just before TDC. The reason for injecting the fuel at 170 bar is so that it can enter into the compressed air and circulate a little before igniting. If the fuel were at

FIGURE 3.9
Air intake control - fuel injection

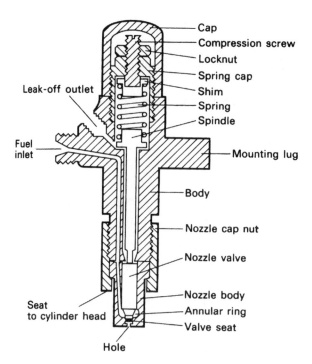

FIGURE 3.10
Diesel injector

Labels in figure:
- Cap
- Compression screw
- Locknut
- Spring cap
- Shim
- Spring
- Spindle
- Leak-off outlet
- Fuel inlet
- Mounting lug
- Body
- Nozzle cap nut
- Nozzle valve
- Nozzle body
- Annular ring
- Valve seat
- Seat to cylinder head
- Hole

a lower pressure than the compressed air it would not be able to enter the combustion chamber.

The injector breaks the fuel into a fine mist of almost atom size particles — atomization. You should remove and test the injectors at regular service intervals — typically every 24 000 miles. The test comprises of looking at the spray pattern made by the injector nozzle and checking the pressure at which the injector opens.

SAFETY NOTES

The injector spray pattern looks very attractive, like a shower head, but do not be tempted to place your hand under the spray — at 170 bar the finely atomized diesel fuel can go straight through your skin and inject a lethal dose into your blood stream.

Venting (also called bleeding)

If you remove any of the diesel fuel system components you will need to vent out the air from the system before you can restart the engine. The fuel filter and the injector pump are both fitted with vent valves, small screws, which when undone, will allow the air to escape.

Basically, you open the vent valves, then operate the fuel pump primer lever manually so that fuel flows from the tank to the injector pump. Close the valves one at a time as the diesel fuel can be seen to squirt out. That is, as the fuel flows to prime the system the air is vented out, the fuel will first flow out of the filter inlet side; tighten that vent valve. Carry on pumping and the flow will move to the outlet side; tighten that valve and then you can move on to the injector pump.

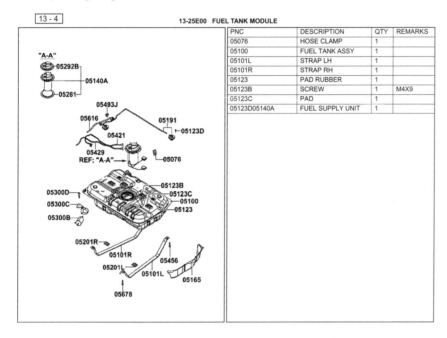

13 - 4   13-25E00 **FUEL TANK MODULE**

PNC	DESCRIPTION	QTY	REMARKS
05076	HOSE CLAMP	1	
05100	FUEL TANK ASSY	1	
05101L	STRAP LH	1	
05101R	STRAP RH	1	
05123	PAD RUBBER	1	
05123B	SCREW	1	M4X9
05123C	PAD	1	
05123D05140A	FUEL SUPPLY UNIT	1	

FIGURE 3.11
Fuel tank module

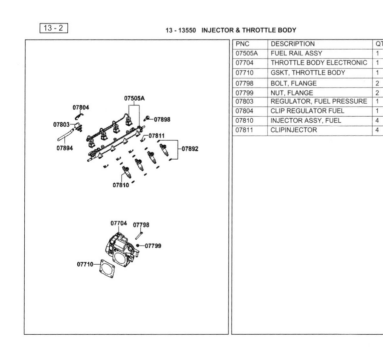

13 - 2   13 - 13550 **INJECTOR & THROTTLE BODY**

PNC	DESCRIPTION	QTY	REMARKS
07505A	FUEL RAIL ASSY	1	
07704	THROTTLE BODY ELECTRONIC	1	
07710	GSKT, THROTTLE BODY	1	
07798	BOLT, FLANGE	2	M6X1X50
07799	NUT, FLANGE	2	M6X1
07803	REGULATOR, FUEL PRESSURE	1	
07804	CLIP REGULATOR FUEL	1	
07810	INJECTOR ASSY, FUEL	4	
07811	CLIPINJECTOR	4	

FIGURE 3.12
Injector and throttle body

MULTIPLE-CHOICE QUESTIONS

1. The petrol tank, for safety reasons, is fitted:
   (a) under the bonnet
   (b) inside the passenger compartment
   (c) in the boot
   (d) outside the passenger compartment

2. The air : petrol ratio at cruising speed is about:
   (a) 11:1
   (b) 20:1
   (c) 15:1
   (d) 17:1

3. The component which pumps diesel fuel at high pressure to the injectors is called:
   (a) injector pump
   (b) lift pump
   (c) air pump
   (d) compressor

4. Air filters must be changed every:
   (a) time they look dirty
   (b) at the service intervals if they look dirty
   (c) at the service intervals whether they look dirty or clean
   (d) time the car is filled with petrol

5. The carburettor needle valve controls the:
   (a) amount of air
   (b) flow of petrol
   (c) mixture strength
   (d) slow running speed

6. The function of the carburettor is to:
   (a) mix petrol and water
   (b) mix air and oil
   (c) mix petrol and air
   (d) mix up the diesel fuel

7. The amount of petrol delivered by a PI system is controlled by the:
   (a) ECU
   (b) EFI
   (c) I
   (d) DERV

8. The injector pipes are:
   (a) made with very thick walls
   (b) made with very thin walls
   (c) made of plastic
   (d) made with push-on ends

9. A typical pressure for a diesel fuel injector to work at is:
   (a) 170 bar
   (b) 2500 bar

(c) 30 bar
(d) 5 bar
10. Diesel fuel vent valves can be found on the:
(a) fuel tank
(b) fuel filter housing
(c) fuel filter element
(d) lift pump

(Answers on page 253.)

FURTHER STUDY

1. Compare the fuel consumption figures of a range of petrol and diesel
vehicles.
2. List the differences between petrol and diesel fuel supply systems.
3. Examine the different types of fuel fillers — petrol and diesel — and report
on how they differ.

Ignition System

The purpose of the ignition system is to provide a **spark** in the combustion chamber of the SI engine which will **ignite the mixture** of petrol and air whilst it is under pressure. As the piston compresses the petrol/air mixture on the compression stroke, the pressure may be increased to over 1000 kPa (150 psi). The voltage needed for the spark to jump across the spark plug gap at this high pressure is about **10 kV** (10 000 volt).

1. THE KETTERING SYSTEM

There are many different types of ignition system; however, the original system, which has been used for about 100 years, is the Kettering System, designed by Dr Kettering. The fully electronic and other systems are easier to understand if you first understand the mechanical–electrical Kettering System first. The main components of the Kettering System are the **battery**, **ignition switch**, **coil**, **spark plugs** and **distributor**. Let's look at these components individually and see how the system works.

RACER NOTE

Lots of kit cars and racing specials use the Kettering System.

Battery

The battery is the **source of electrical power** for the ignition system (and other systems too), it is usually located in the engine compartment. The

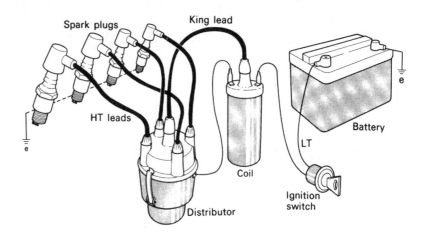

FIGURE 4.1
Ignition system layout

battery supplies electrical power to the ignition switch. The battery has a nominal voltage of **12 volts**; the ignition circuit draws about 0.5 amp. The battery is usually connected negative side (−) to the chassis earth and positive side (+) to the main **fuse box**.

RACER NOTE

Wiring on race cars and bikes is usually through a separate isolator switch, and may incorporate a connection for a trolley-mounted starter battery, or power pack.

SAFETY NOTE

Battery acid is corrosive — mind that you do not spill it.

Ignition switch

This is connected to the battery through the main fuse box. The ignition switch makes and breaks the ignition circuit. When the engine is not running, the switch must be off to disconnect the electrical power supply and so prevent the coil from overheating. Switching the ignition off also switches off other related circuits and prevents the battery from being **discharged**. The ignition switch is also a **security device** to prevent the vehicle from being stolen or used unlawfully. The ignition switch is combined with the steering lock for added security, and for the convenience of needing only one key for both the steering lock and the ignition switch. The ignition switch/steering lock is situated on the side

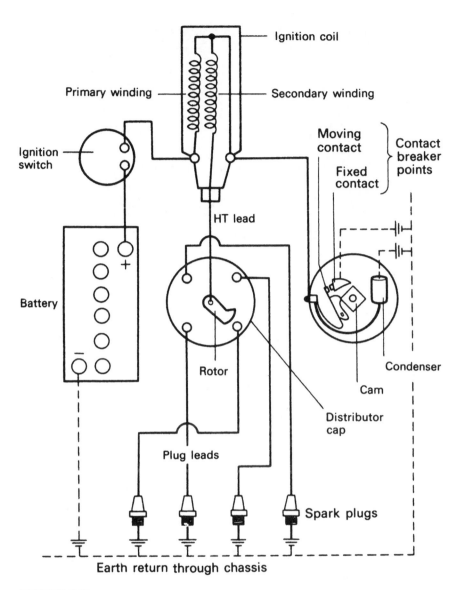

FIGURE 4.2
Diagrammatic layout of ignition

of the steering column near to the dashboard within easy reach of the driver. The ignition switch on road cars also incorporates the starter switch.

RACER NOTE

Race cars usually have separate ignition switch and starter buttons which are covered when not in use to prevent accidental operation.

FIGURE 4.3
Scrutineering label

Ignition coil

The ignition coil is a kind of transformer. It changes the **low tension (LT)** 12 V from the battery to a **high tension (HT)** 10 kV at the spark plugs.

Nomenclature

Low tension (LT) refers to the components of the ignition system, which operate at a nominal voltage of 12 V. High tension (HT) is that part of the ignition system which operates at several thousand volts. The HT can range from 5 kV for an older car, to 40 kV for a motorsport vehicle. We refer to the battery has having a nominal voltage of 12 V. We say this is a 12 V battery, but the actual voltage may be between 9 V and 16 V depending on a number of factors. A kilovolt is 1000 volts.

The coil has three electrical terminals, one that goes to the ignition switch, one to the distributor body and one to the distributor cap. The ones to the ignition switch and the distributor body are low tension; the one to the distributor cap is high tension.

SAFETY NOTE

The high voltage spark at the spark plugs, and associated leads, can kill you. If the system is wet, or you touch a bare cable or component, the shock will cause you to react and you may hit your head, or elbow, on another part of the car which could result in a serious injury.

The connection to the ignition switch is the 12 V power supply. The connection to the distributor body goes to the contact breaker points, which carries out the switching action. The HT to the distributor provides the spark for the spark plugs.

The ignition coil operates on the principle of **difference of turns**. That is, it has two separate **windings** wound around a **soft iron core**. The **primary winding** is connected to the distributor LT; the **secondary**

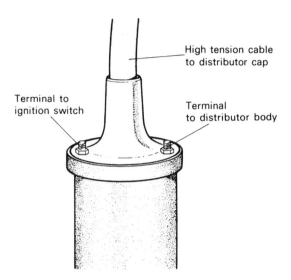

High tension cable
to distributor cap

Terminal to
ignition switch

Terminal
to distributor body

FIGURE 4.4
HT coil

winding is attached to the distributor HT. The secondary winding has many more turns of wire than the primary winding; the increase in voltage from LT to HT is proportional to the difference in the number of turns of the wire.

Ignition coils are usually filled with oil to improve cooling. Be careful not to damage the coil case as this is usually made from soft aluminium, which if damaged may allow the oil to drain and the coil to overheat.

61

Spark plugs

The metal end, or **body**, of the spark plug screws into the cylinder head so that the **electrode** protrudes into the combustion chamber. The screw thread with the washer and mating surfaces form a gas-tight seal, so prevent the loss of compression from the cylinder.

The spark plug has two electrodes, a **centre electrode** and a **side electrode**. The terminal end unscrews for use on motorcycles and agricultural machines, but on cars the **plug leads** are usually a push-fit.

The diameter and the reach of the spark plug vary from engine to engine. The most common diameters are 10 mm, 14 mm and 18 mm. The common reaches are 3/8 inch, 1/2 inch and 3/4 inch. It is important that the correct reach of spark plug is fitted to the engine, otherwise the electrode may foul against the piston crown. The diameter will only fit the tapping size in the cylinder head. The spark plug can be identified by the letter and number code that is printed on either the **insulator** or the body.

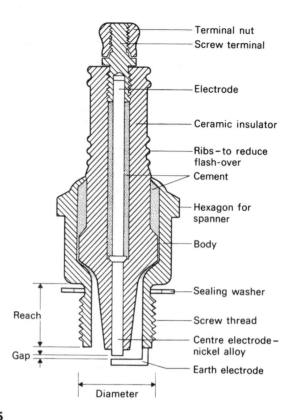

- Terminal nut
- Screw terminal
- Electrode
- Ceramic insulator
- Ribs – to reduce flash-over
- Cement
- Hexagon for spanner
- Body
- Sealing washer
- Screw thread
- Centre electrode – nickel alloy
- Earth electrode

Reach

Gap

Diameter

FIGURE 4.5
Spark plug

62

FAQs

How can I check that one make of spark plug matches another?
Spark plugs are made by a number of different companies, for example: Champion, Bosch and NGK. Most motor parts and accessory shops have lists of equivalents. These lists show the various makes and model of cars and tabulate the code numbers for the different spark plug manufacturers.

When spark plugs are replaced, typically at between 16 000 km (10 000 miles) for old vehicles and 60 000 km (40 000 miles) for modern vehicles, they must be replaced with the correctly coded ones. The spark plug manufacturer's list, or the workshop manual, should be used to check the exact code for the model. It is common for models of different specification or year to use different spark plugs.

Servicing spark plugs is limited to cleaning and gapping between replacements. If cleaning is needed then a special machine is used. Gapping the plug means setting the size of the **gap** between the fixed and the side electrode.

Spark plug gaps are usually between 0.020 and 0.040 inch (0.5 and 1 mm). New spark plugs are usually ready-gapped from the factory, but it is worth checking them before fitting. Plug condition is usually indicated by the engine analyser test.

RACER NOTE

Setting spark plug gaps takes skill and practice. Hold the feeler gauge on the flat surfaces between your first finger and thumb. The correct clearance can be felt as a faint touching drag when you move the feeler gauge through the gap. Close the gap by gently tapping the side electrode on a hard metal surface such as a bench top. Do not use a screwdriver or pliers to alter the gap. Before fitting new spark plugs always check the gap settings. When tightening up spark plugs always use a correctly set torque wrench. The taper fitting spark plugs must not be over-tightened. If the spark plug thread in the cylinder head is slightly rusty, or, in an aluminium cylinder head, dry, apply a little light oil to the plug threads before fitting them.

Distributor

The inside of the distributor is divided into three parts. The lower part houses the mechanical components and the linkages, the middle part comprises the LT components and the upper part is the HT section. The HT components are mainly the **distributor cap** and the **rotor arm**.

HT electricity is delivered from the coil to the centre of the cap. Electricity flows from the cap to the rotor arm through a **brush** arrangement. The **rotor arm** is connected to the **distributor spindle** so that it goes around at the same speed as the **spindle**. As the rotor arm rotates, its free end aligns with **segments** in the cap, one at a time. The sequence is the firing order. As each segment is aligned, the current is passed from the rotor arm to the segment.

63

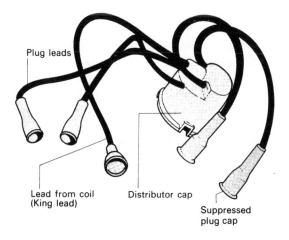

Plug leads

Lead from coil
(King lead)

Distributor cap

Suppressed
plug cap

FIGURE 4.6
HT cap and leads

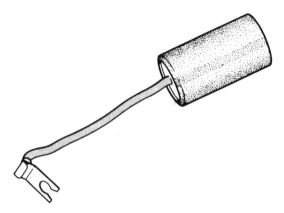

FIGURE 4.7
Capacitor (also called condenser)

The segments each have a **plug lead** attached to carry the electricity on its way to the spark plug.

The low tension part of the distributor comprises the **contact breaker (cb) points** and the **capacitor**. The **cam ring** which is formed on the outside of the distributor spindle rotates at the same speed as the spindle. As the cam goes round, it opens and closes the cb points. It is this action which causes the current to flow in the HT circuit by **induction** in the coil. The **gap** of the cb points, and their position in relation to the spindle, affects the **ignition timing** and general efficiency. The cb points should be checked for condition and size of gap every 8000 km (5000 miles). They should be replaced every 16 000 km (10 000 miles).

The **capacitor** (also called a **condenser**) is fitted to give a good quality spark by controlling the flow of electricity; this reduces **arcing** at the points and gives a longer life to the cb points.

The drive for the distributor is picked up from the camshaft by an angular gear called a **skew gear**. This skew gear is on the lower end of the distributor spindle. The spindle passes through the mechanical and the vacuum timing mechanisms. These mechanisms advance and retard the ignition timing to suit different engine speeds and load characteristics.

FAQs

Does the distributor rotate at the same speed as the crankshaft, or that of the camshaft?

The distributor rotates at the same speed as the camshaft, that is, half the speed of the crankshaft. Therefore, on a four cylinder engine the cb points open four times for each revolution of the distributor spindle.

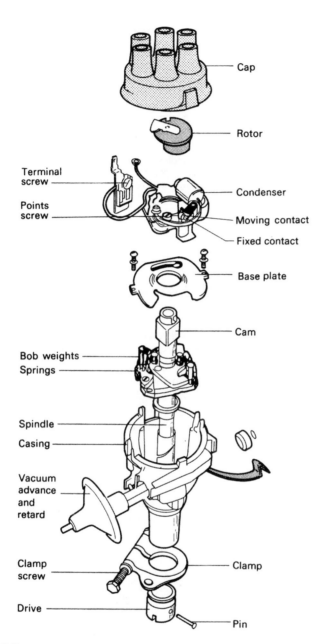

Cap

Rotor

Terminal screw

Condenser

Points screw

Moving contact

Fixed contact

Base plate

Cam

Bob weights

Springs

Spindle

Casing

Vacuum advance and retard

Clamp screw

Clamp

Drive

Pin

FIGURE 4.8
Distributor

65

Ignition timing

The ignition system is designed so that the spark occurs in the combustion chamber a small number of degrees **before top dead centre** (BTDC). Car manufacturers give specific figures for each of their different models. Typically, the static timing is 10 degrees BTDC — that is, when the engine is rotated by hand. When the engine is running, dynamic timing, the timing will advance to about 30 degrees BTDC.

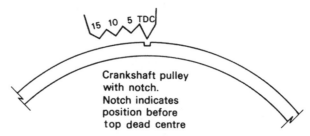

Pointer attached to block marked in degrees before top dead centre

15 10 5 TDC

Crankshaft pulley with notch. Notch indicates position before top dead centre

FIGURE 4.9
Timing marks

RACER NOTE

You will find the ignition timing settings in the workshop manual or service data book. The static timing can be easily checked by using a small low wattage bulb (such as a side lamp bulb) in a holder with two wires attached. Attach one wire to the distributor terminal on the ignition coil — usually marked with a positive sign (+). Attach the other wire to a good chassis earth. Disconnect the HT king lead, or remove the distributor cap so that the engine will not start. Switch on the ignition and turn the engine by hand. The bulb should light just as the timing mark comes into line. The timing can be adjusted by slackening the distributor clamp and moving the distributor gently. With the timing marks aligned, move the distributor until the bulb just lights.

Dwell angle

This is the period for which the contact breaker points remain closed. When the cb points are closed, the magnetic field builds up in the ignition coil.

FIGURE 4.10
ECU

When the points open, the spark is triggered at the plugs. The dwell angle is directly proportional to the number of cylinders and the cb points gap and as such is used as an indicator to the engine condition. For four-cylinder engines it is typically 60 degrees.

2. ELECTRONIC IGNITION

There are many different types of electronic ignition. The most common type has a distributor which does not have cb points. The distributor has an electronic trigger device instead of the cb points. On the outside of the distributor is an amplifier unit. The switch, coil and battery remain the same.

There are three main types of trigger devices used instead of cb points. These are:

- **Inductive type** – this uses a magnet attached to the distributor shaft to induce a small electrical current into a pick-up coil. The current produced is low voltage (typically 2 V) alternating current (AC).
- **Hall effect** – this uses a small integrated circuit (IC) as a switch which is turned on and off by the passing of a metal drum-shaped component. This works at low voltage (typically 5 V).
- **Optical type** – this uses an infrared light emitting diode (LED) which shines on a phototransistor. The light is turned on and off by a Maltese Cross-shaped component usually referred to as a light chopper. Operating voltage is typically 9 V.

As the electronic ignition system does not have cb points there are less moving parts, so reliability engine economy and performance are all improved. The only ignition servicing that is required is changing the spark plugs. On electronic ignition cars the spark plugs may only need changing every 60 000 km (40 000 miles)

Other types of electronic ignition do not have a distributor and do not have a conventional coil. Also, the ignition lock and key have become a more complicated security device. The key incorporates a very small electronic device called a transponder; this transponder is a sort of electronic key that electronically unlocks the ignition at the same time that it is being mechanically unlocked.

FIGURE 4.11
Coil pack

3. DISTRIBUTOR-LESS IGNITION SYSTEM (DIS)

The distributor-less ignition system (DIS) is usually part of an engine management system which has two parts: **ignition** and **fuel**. Looking at the ignition part, DIS comprises:

- **ignition IC** (chip) in the Electronic Control Unit (ECU)
- **sensors** for: **crankshaft position**, TDC, **knock** (pinking), **throttle position**, **engine temperature**
- **spark plugs**
- **ignition coils** for each spark plug.

RACER NOTE

Sensors on different makes of systems may vary and have different values for different engines.

The ignition IC is **programmed** to trigger a spark according to engine load and road conditions — it is fully self-regulating and needs no servicing, apart from spark plug replacement at set intervals. In the event of a fault, this is found using **electronic diagnostic equipment.**

16 - 5   16-16220 SPARK PLUG, CABLE AND COIL

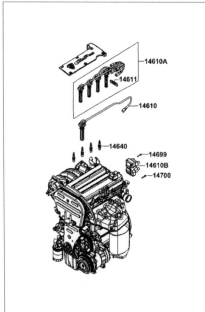

PNC	DESCRIPTION	QTY	REMARKS
14610A	ASSY CABLE SET SPARK PLUG	1	
14610	CABLE SPARK PLUG	1	
146108	IGNITION COIL	1	
14611	CLIP CABLE SET	1	
14640	SPARK PLUG	4	
14699	BOLT FLANGE	2	M5X0.8
14700	NUT FLANGE	2	M5XO.8

FIGURE 4.12
Spark plug cable and coil

MULTIPLE-CHOICE QUESTIONS

1. The source of electrical power for the ignition system is the:
   (a) coil
   (b) battery

    (c) distributor

    (d) starter motor

2. The ignition coil gives an output of:

    (a) 1 kV

    (b) 10 kV

    (c) 100 kV

    (d) 1000 kV

3. Electrodes, terminal and body are all found on a:

    (a) distributor

    (b) spark plug

    (c) capacitor

    (d) ignition switch

4. The spark occurs at the spark plugs just:

    (a) ATDC

    (b) BTDC

    (c) ABDC

    (d) BBDC

5. The component which prevents arcing at the cb points is the:

    (a) coil

    (b) capacitor

    (c) king lead

    (d) timing lamp

6. The component in the ignition system which is designed to prevent the theft of the car is the:

    (a) ignition switch

    (b) rotor arm

    (c) earth lead

    (d) battery

7. An amplifier and a transponder are likely to be found in which ignition system:

    (a) Kettering

    (b) Electronic

8. HT voltage may vary between 5 kV and 40 kV:

    (a) True

    (b) False

9. The distributor is driven by a skew gear from the:

    (a) crankshaft

    (b) camshaft

    (c) timing chain

    (d) fan belt

10. On a four-cylinder distributor, one revolution of the spindle will open the cb points:

    (a) once

    (b) twice

    (c) four times

    (d) eight times

(Answers on page 253.)

69

FURTHER STUDY

1. Ignition system components cause many vehicle breakdowns, so it is important to be able to actually remove and refit all the components that we have discussed in this chapter. On a running engine, either in or out of a car, remove the ignition components and refit them using a workshop manual for guidance.

2. Investigate an electronic ignition system on a car of your choice and sketch the wiring connections — use a workshop manual for help.

3. Locate a spark plug identification table and identify the spark plugs that may be used in six vehicles of your choice.

CHAPTER 5

The Cooling System

KEY POINTS

- The purpose of the cooling system is to keep the engine at a constant temperature, whilst preventing the overheating of any of the individual components.
- Typically car petrol engines run at between 80 and 90 degrees Celsius (180 and 190 degrees Fahrenheit), this is the best temperature to get the best fuel consumption and the least pollution. Diesel engines run at about 5 degrees Celsius (10 degrees Fahrenheit) cooler.
- There are two main types of cooling system: liquid (water) cooling and air cooling.

1. LIQUID (WATER) COOLING SYSTEM

FAQs

Is water cooling the same as liquid cooling?

Yes, the liquid in the cooling system is water mixed with a number of chemicals and is sometimes referred to by mechanics simply as water. Because it is not just water, its proper name is coolant.

SAFETY NOTE

Coolant in an engine is likely to be scolding hot and under high pressure.

- Do not touch cooling system components or remove filler caps when hot.

The liquid cooling system works by using **coolant**, the name for water mixed with other chemicals, to take the heat from the cylinder block and pass it to the radiator so that it is cooled down. That is, the coolant circulates through the

Basic Motorsport Engineering.

engine, where it gets hot, then goes through hoses to the radiator where it cools down again, then finally, back through another hose to the engine to go through the process again. When the petrol or diesel fuel is burning in the combustion chamber the temperature gets very hot — about 2000 degrees Celsius (3500 degrees Fahrenheit). The cooling system therefore needs to work very hard to keep the temperature of the components at about **85 degrees Celsius (185 degrees Fahrenheit)**. Most engines run at between about 80 and 90 degrees Celsius (180 and 190 degrees Fahrenheit), diesel engines tend to run about 5 degrees Celsius (10 degrees Fahrenheit) cooler than petrol engines. The temperature of the engine is kept between 80 and 90 degrees Celsius (180 and 190 degrees Fahrenheit) because this is its most efficient temperature, that is to say it will use the least fuel and produce the least pollution.

Coolant has a natural tendency to circulate when heated up in the cooling system. The hot liquid rises, the cooling liquid falls. This is called **thermo-siphon**. As the coolant is heated up in the cylinder block water jackets it rises up; it then passes through the **thermostat** into the **top hose** to the radiator **header tank**. The water then falls through the **radiator core** into the **bottom tank**. As the coolant falls inside the radiator core, it is cooled by the incoming air which passes around the outside of the **radiator fins**. The incoming air is from the front of the vehicle and may be forced along by the fan. The weight of the water in the radiator forces the coolant through the **bottom hose** back into the engine. The **coolant (water) pump** helps the coolant circulate more quickly into the water jackets. For the coolant to be able to circulate, the water level must be kept above the top hose connection so that it can maintain a flow into the radiator.

The coolant

The liquid used in a cooling system is usually called the coolant, because it cools the engine. The coolant is a mixture of **water**, **anti-freeze** and a chemical which **inhibits corrosion** to the metal parts inside the engine. As you probably know, water boils at 100 degrees Celsius (212 degrees Fahrenheit) and freezes at 0 degrees Celsius (32 degrees Fahrenheit). This means that in winter water could freeze and damage the engine. Typically the coolant mixture boils at about 110 degrees Celsius (230 degrees Fahrenheit) and freezes at about minus (−) 18 degrees Celsius (0 degrees Fahrenheit).

FAQs

What is anti-freeze made from?
It is a chemical called ethylene glycol.

Coolant can be bought ready-mixed, which is usually advised for specialist engines, or for most other vehicles, anti-freeze, which contains anti-corrosion chemicals, can be mixed with water. For British winters it is normal to mix a 33 per cent anti-freeze solution, that is 1/3 anti-freeze and 2/3 water: the solution must be measured accurately.

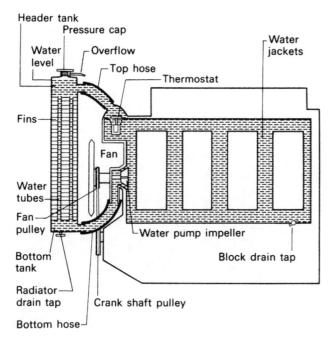

FIGURE 5.1
Cooling system layout

The strength of the coolant can be checked with a special hydrometer called an anti-freeze tester; usually this test must be carried out when the coolant is cold.

Key Points

When mixing anti-freeze:

- always use clean water to prevent damage to the engine
- check the instructions on the label
- mix the anti-freeze and water before putting it in the engine.

RACER NOTE

Coolant is one of the life bloods of an engine (the other is the oil). So always use the most appropriate coolant for your engine. Look for coolant which contains a wetting agent as well as a corrosion inhibitor and anti-freeze. Aluminium engines are prone to internal corrosion if left only partially drained.

Key Points

Anti-freeze loses it properties after about two years, so change it every two years, or at the major services as recommended by the vehicle manufacturer.

FIGURE 5.2
Custom shop — Caribbean style

> ## RACER NOTE
>
> Coolant temperatures will vary with race conditions; however, whether you win or not you should keep the Champagne in the fridge between 4 and 6 degrees Celsius (39 to 42 degrees Fahrenheit).

Coolant (water) pump

To circulate the coolant quickly a pump is fitted; this is called a coolant or water pump. The coolant pump is fitted to the front of the engine. The bottom hose from the radiator is fitted to the coolant pump inlet so that it is supplied with cooled coolant; the pump outlet is connected directly to the water jacket of the cylinder block.

The coolant pump is driven either by the **fan belt**, or the **cam belt**, depending on the engine design.

Fan

The fan is used to draw in air through the grille or ducting and pass it over the outside of the **radiator fins** so that it cools the coolant inside the radiator core.

The fan may be of the mechanical or the electrical type. The mechanical type is usually attached to the water pump spindle so that it is also driven by the fan belt. The fan will be running all the time that the engine is running. The electrical type is operated by an electric motor, which is only switched on

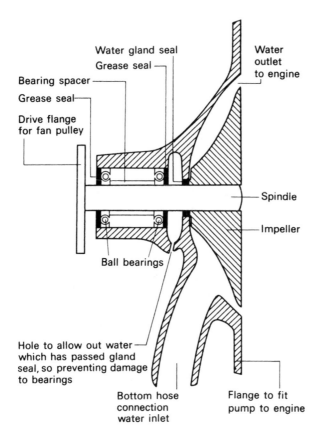

FIGURE 5.3
Coolant (water) pump

when the engine gets hot. There is a temperature-sensitive switch mounted on the radiator, so that when the radiator gets to a preset temperature the fan is switched on, when it cools down it is switched off.

Radiator

The radiator is made up of the **header tank**, the **bottom tank** and the **core** between them. The flattened tube type of construction is used on most vehicles; heavy goods vehicle tend to use the round tube variety; the honeycomb design is only used on a few luxury or high performance cars. The radiator fins are there to increase the surface area of the cooling zone. That is, the fins dissipate, or spread out, the heat more efficiently so that it is transferred to the air quickly.

The radiator may be either **vertical** in design, with top and bottom tanks stacked with the core, or of **cross-flow** design. The cross-flow radiator has its tanks mounted at each end; this means that the coolant flows from one side of the radiator to the other, rather than from the top to the bottom. The cross-flow radiator is used to give low bonnet height.

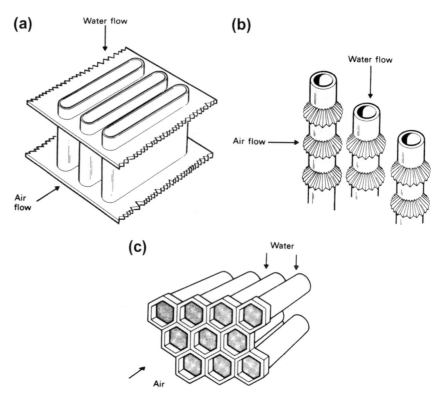

(a) Flattened tube; (b) round tube; (c) honeycomb

FIGURE 5.4
Radiator core types

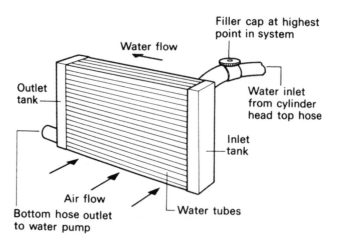

FIGURE 5.5
Cross-flow radiator

Thermostat

The thermostat is fitted in the cylinder head; it is a sort of tap to control the coolant flow between the engine and the radiator. The thermostat is usually in a special connector, one end of this holds the thermostat against the inside of the cylinder head, and the other end is attached to the top hose.

When the thermostat is closed the coolant cannot flow; when it is open it can flow freely. The thermostat allows a short warm-up period by remaining closed until the engine has reached its required temperature. It keeps the engine at a constant temperature by opening and closing as the engine becomes hot or cools down.

The thermostat shown in the diagram is a wax-stat.

FAQs

Why is it called a wax-stat?
Because it is a **wax** operated thermo**stat**.

The metal body or capsule is filled with wax, which expands in volume very rapidly at the temperature at which its designers want it to open — usually around 80 degrees Celsius (180 degrees Fahrenheit). So when the wax reaches the design temperature it rapidly expands, forcing the thrust pin out

77

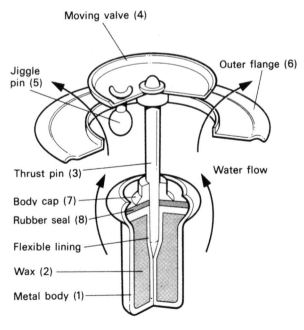

FIGURE 5.6
Thermostat

of the capsule against the pressure of the return spring (not shown). The thrust pin lifts the moving valve to allow coolant to flow from the engine to the radiator through the top hose. When the wax cools it contracts, the return spring closes the valve and returns the thrust pin into the capsule. The jiggle pin is fitted to prevent the formation of a vacuum in the engine water jacket by allowing small amounts of coolant to flow even when the thermostat is closed.

Key Points

You'll find the opening temperature of the thermostat stamped on either the rim or the capsule as a two digit number.

Radiator pressure cap

SAFETY NOTE

Never remove the radiator cap when the engine is HOT or RUNNING.

At normal atmospheric pressure, water boils at 100 degrees Celsius (212 degrees Fahrenheit). At high altitudes the boiling temperature is reduced. This also applies to the coolant mixture, though the temperatures may be slightly different. To prevent the engine from boiling and over-heating, a radiator pressure cap is fitted. The force of the spring in the cap ensures that the coolant is kept under pressure. The higher the pressure, the higher the boiling point of the coolant in the system. To prevent the build-up of a vacuum inside the radiator, a vacuum valve and spring is fitted. The vacuum valve prevents the radiator from imploding, or collapsing inwards, when the coolant temperature is reduced and the coolant pressure therefore decreases.

Key Points

Radiator caps are made to operate at different pressures — the design pressure is usually written on the outside of the cap — this may be in pounds per square inch (psi) or bars.

The actual retaining cap is secured to the radiator neck with two curved sections. The radiator cap is fitted to the radiator in a similar way to the lid on a jam jar. The pressure seal is held in place against the top of the radiator by the pressure spring. Only when the coolant pressure exceeds the preset figure — which is stamped on the radiator cap — is the seal lifted against spring pressure. The coolant is then released through the overflow pipe. When the pressure of the coolant in the radiator is released by the coolant running off, the spring will press the seal back onto the top of the radiator. When the radiator cools, its coolant pressure decreases. Air pressure from

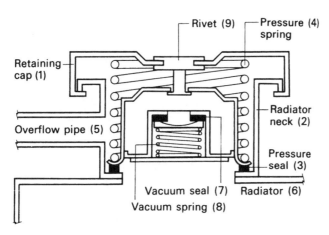

FIGURE 5.7
Radiator (coolant) pressure cap

outside the radiator forces the vacuum seal against the force of the vacuum spring to allow air to pass over the seal into the radiator.

Failure of the radiator pressure cap causes overheating and boiling over. The operating pressure of the cap can be tested with a special pressure gauge and the condition of the seals can be inspected visually.

Sealed cooling system

To prevent the loss of coolant, and hence the need for topping up the radiator, most vehicles use sealed cooling systems. An **overflow tank** is fitted to the side of the radiator – sometimes this can be a little way from the radiator on the inner wing panel. A tube from the radiator neck is connected to the overflow tank. Therefore any coolant which is allowed passed the radiator cap will go into the overflow tank. The tube is arranged so that its end is always below the coolant level – see Figure 5.8 – therefore, when the radiator cools down and the coolant contracts, the coolant in the overflow tank is drawn back into the radiator to fill the available space. The overflow tank must be kept partially full of coolant to ensure that coolant covers the bottom of the tube.

Hoses

Rubber hoses are used to connect the engine to the radiator so that there is some flexibility between the two components. The engine is free to move slightly on its rubber mountings, but the radiator is rigidly mounted to the vehicle's body/chassis. The flexing of the hoses causes them to deteriorate; they usually crack, and if not replaced they will break or puncture causing the total loss of coolant. The hoses are held in place with clips – the clips usually have a screw mechanism to tighten them up.

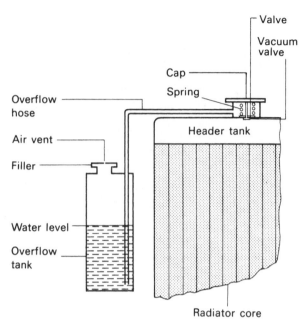

FIGURE 5.8
Sealed cooling system

FIGURE 5.9
Cruise without booze

Hose clip drivers are available with hexagonal ends, which are much safer to use and less likely to slip and cause damage to the radiator than a screwdriver.

When you are fitting a new hose always check that the surfaces are clean, and if possible use a new clip.

2. AIR COOLING

Air cooling is used on most motorcycles, a few specialist cars − like old Volkswagens and Porsches − and some specialist agricultural machinery.

Air cooling has the advantage of not using a liquid coolant (water) and having fewer moving parts. Having no water it cannot freeze or leak. However, air-cooled engines tend to be more noisy than liquid cooled ones. The air cooling system operates by air entering through the flap valve. The fan, which is driven by the crankshaft pulley, forces the air over the fins on the cylinders. The air is then discharged back into the atmosphere. The flap valve is controlled by the thermostat, which opens when the engine is hot, so allowing air to enter. The flap valve is closed when the engine is cold; this restricts the air flow to give the engine a quick warm-up to operating temperature.

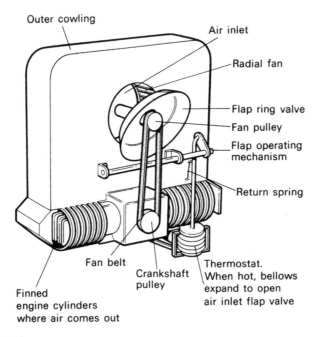

FIGURE 5.10
Air cooling systems − as used on classic Volkswagens (VW)

Fan belt

A V-shaped fan belt, sometimes called a V belt, is used to drive the fan, the water pump and the alternator. It is important that the belt is free from cracks and shredding. It must be adjusted to give between 10 and 20 mm (1/2 to 3/4 inch) of free-play on its longest side. The fan belt adjustment is usually carried out by moving the alternator on its slotted elongated mounting bracket. The alternator bracket is usually attached to the cylinder block at one end, and given an elongated slot at the other where the alternator is attached. To check the adjustment you should pull and push the fan belt with your finger and thumb on the longest section. Be sure when you are doing this that the engine is switched off and the key is removed.

Some water pumps are driven by the cam belt; and some fans by electric motors.

SAFETY NOTE

Do not work on a running engine; keep your tie and jewellery away from the fan belt area.

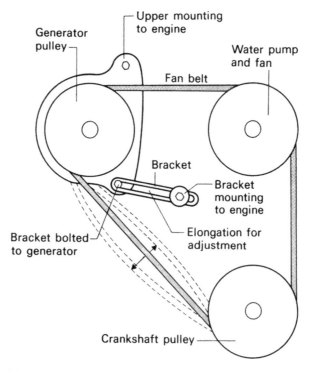

FIGURE 5.11
V-belt adjustment

14-14110 WATER PUMP

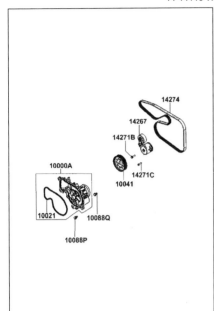

PNC	DESCRIPTION	QTY	REMARKS
10000A	ASSY WATER PUMP	1	
10021	SEAL WATER PUMP TO FR CASE	1	
10041	PULLEY WATER PUMP	1	
10088P	BOLT, FLANGE	8	M6X1X16
10088Q	SCREW SPECIAL	6	M6X1X12
14267	ASSY TENSIONER ACC DRIVE	1	
14274	BELT ACC DRIVE	1	
14271B	BOLT FLANGE	1	M10X1.25X90
14271C	BOLT FLANGE	1	M8X1.25X30

FIGURE 5.12
Water pump

14-25CO1 RADIATOR ASSY

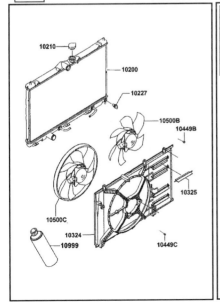

PNC	DESCRIPTION	QTY	REMARKS
10200	RADIATOR ASSY	1	
10210	CAP	1	
10227	COCK SUB ASSY	1	
10324	SHROUD ASSY, FAN	1	
10325	DOOR FLAP	1	

FIGURE 5.13
Radiator assembly

MULTIPLE-CHOICE QUESTIONS

1. A typical running temperature for a car cooling system is:
   (a) 0 degrees Celsius
   (b) 32 degrees Celsius
   (c) 85 degrees Celsius
   (d) 100 degrees Celsius

2. The cooling system component which gives a quick warm-up and keeps the engine at a constant temperature is the:
   (a) radiator
   (b) water pump
   (c) pressure cap
   (d) thermostat

3. A cross-flow radiator is almost certainly to be found in a:
   (a) truck
   (b) bus
   (c) limousine
   (d) sports car

4. Two advantages of an air cooling system are it:
   (a) is cheap and easy to clean
   (b) uses no water and has few moving parts
   (c) is relatively quiet and does not freeze up
   (d) is cheap and quiet

5. The component which circulates the coolant is called the:
   (a) radiator
   (b) distributor
   (c) water pump
   (d) thermostat

6. To prevent the radiator from imploding when cooling, the radiator cap is fitted with a:
   (a) pressure valve
   (b) vacuum valve
   (c) water jacket
   (d) small screw

7. Flexible rubber hoses are fitted to allow:
   (a) movement between the engine and the radiator
   (b) high operating temperature
   (c) low operating temperatures
   (d) the use of anti-freeze

8. A fan belt can best be described as being shaped like the letter:
   (a) Z
   (b) V
   (c) Y
   (d) X

9. A 33 per cent anti-freeze solution means that one part of anti-freeze is added to how many parts of water?
   (a) 1
   (b) 2
   (c) 3
   (d) 4
10. The typical free-play in a fan belt is:
   (a) 5 mm
   (b) 10 mm
   (c) 50 mm
   (d) 100 mm

(Answers on page 253.)

FURTHER STUDY

1. Using the workshop manual for a car of your choice, find out the operating temperature and pressure of the cooling system.
2. On a vehicle of your choice, describe step by step how to check the coolant level.
3. Why is a small amount of free-play needed in a fan belt?

Lubrication System

Lubrication is needed to keep mating bearing surfaces apart, which reduces friction and wear. The lubricant also acts as a coolant, taking heat away from the bearing surfaces to maintain a constant running temperature. The liquid cooling system runs at about 85 degrees Celsius; the lubrication system runs at a slightly higher temperature, 90 to 120 degrees Celsius being typical figures for road use; for racing engines this may exceed 200 degrees Celsius. The lubricant also picks up small particles of metal and carbon from the components that it passes over and deposits them in the oil filter. It is the particles that are too small to be filtered out that make the oil a dirty black colour.

RACER NOTE

Oil temperature gauges are typically red lined at about 150 degrees Celsius.

1. FRICTION

Friction is the resistance of one surface when sliding over another. The amount of friction is referred to as the coefficient of friction and indicated by the Greek letter mu that has the symbol μ.

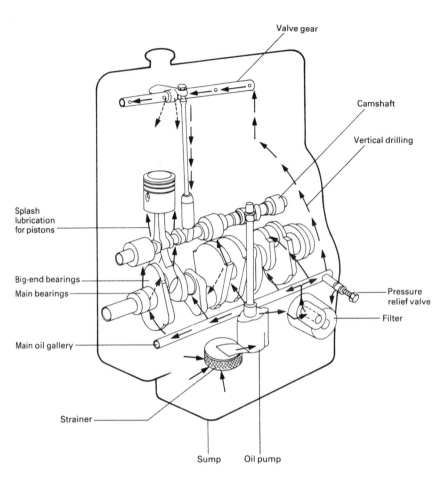

Valve gear

Camshaft

Vertical drilling

Splash
lubrication
for pistons

Big-end bearings

Main bearings

Pressure
relief valve

Filter

Main oil gallery

Strainer

Sump   Oil pump

FIGURE 6.1
Lubrication system layout

To calculate the coefficient of friction the formula used is:

$$\mu = F/W$$

Where F is the force that is required to slide the object over the mating surface and W is the weight of the object. Both F and W are given in the same units, usually Newtons (N). The value of μ is always less than unity; that is between 0 and 1.

Lubrication keeps the mating surfaces apart so that they can slide easily over each other. The simplest form of lubricant is water – think of how slippery a floor is when it is wet. Motor vehicles use a variety of oils and greases in a range of different ways, the following sections looks at some of them.

Nomenclature
μ almost always as a value of less than unity (in other words one) as it is usually easier to push something along than to lift it.

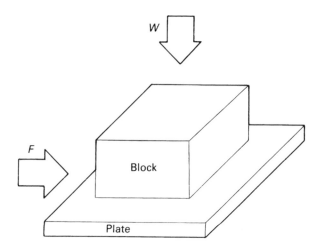

FIGURE 6.2
Friction

2. TYPES OF LUBRICATION

Lubricants are generally used in the form of oil, which is liquid, or grease, which is semi-solid. Oil in the engine is fed to the bearings under pressure; the film of oil keeps the bearing surfaces apart; this is known as **full-film lubrication**. Grease, which is used for steering and suspension joints, and wheel bearings, does not keep the bearing surfaces fully separated; this type of lubrication is known as **boundary lubrication**.

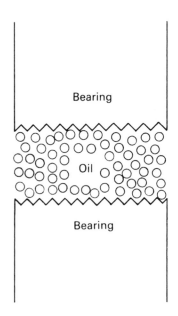

FIGURE 6.3
Full-film lubrication

3. VISCOSITY

Oils are classified according to their viscosity. **Viscosity is the resistance of an oil to flow**, it is calculated by timing how long it takes a fixed quantity of oil to flow through a specific diameter aperture, at a preset temperature. The aperture is a hole in a piece of metal that looks like a washer. The time is in seconds; because the timing is done in this specific way, the seconds are called Redwood Seconds. The longer the amount of time taken, the higher the oil's viscosity is said to be, and the higher the viscosity number it is given. In common language it is said to be thicker. Basic oils are:

5 SAE — Cycle oil
30 SAE — Straight engine oil (not commonly used)
90 SAE — Gear/axle oil
140 SAE — Truck axle oil

SAE is the abbreviation for the **Society of Automotive Engineers**; this is an American-based international organization, which sets standard for many areas of automotive engineering.

When the temperature of oil is raised its viscosity usually decreases; that is to say when it gets hotter it gets thinner too. Therefore most modern oils for vehicle engines are multigrade. That is to say they have two viscosity ratings, one for when they are cold and one for when they are hot. A typical oil is 15/40 SAE. This oil is rated as a thin 15 SAE when it is cold to give easy cold starting; it is rated a thick 40 SAE when it is hot to give good protection to the engine when it is hot — such as on the motorway. Other popular grades are 10/40 SAE and 20/50 SAE.

4. TYPES OF OIL

Oil can be classified in a number of different ways, and it is important to choose the correct oil for the vehicle. Let's look at some of the common types:

- **Mineral oils** are those pumped from the ground, and **vegetable oils** are those, such as **Castrol R** oil, that are made from vegetable products. Castrol R oil is used for a small number of specialist applications such as racing motorcycle engines. Most cars and trucks use mineral oils. Mineral and vegetable oils cannot be mixed.
- **Synthetic oils** are chemical engineered oils; they are specially prepared mineral oils. Synthetic oils are very expensive. A cheaper alternative is **semi-synthetic** oil.
- Oil for diesel engines and ones with turbochargers are specially classified to withstand the very high temperatures and pressures of these engines — look for the marking on the container. Synthetic and other oils may be used for these special purposes.
- **Extreme pressure (EP)** oils are used in gearboxes; usually their viscosity rating is 80 or 90 SAE. They maintain a film of oil under a very heavy load between two gear teeth.

- **Hypoy** (or hypoid) oil is used for a special shape of gear teeth in rear axles on large vehicles and trucks. It is usually of 90 SAE or 140 SAE viscosity.
- **Automatic transmission fluid** (**ATQ** or similar) is an oil which is also a hydraulic fluid. It is used in power steering as well as the automatic gearbox.
- 3-in-1 oil, or **cycle oil**, is 5 SAE and is used because it cleans and prevents rusting, as well as lubricating. On motor vehicles it is used for control cables, door hinges and electrical components.

RACER NOTE

Because of the exacting nature of lubrication in race engines it is essential to use the right oil irrespective of cost — a race engine rebuild may cost more than the price of a small family car. Many of the specialist race lubrication suppliers and manufacturers sponsor race car teams because they enjoy the sport — not just to sell more oil.

5. ENGINE LUBRICATION SYSTEM

The most popular type of engine lubrication is the **wet sump** type. This is so-called because the sump is kept wet by the engine oil, which uses the sump as a supply reservoir.

In operation the oil is drawn from the sump by the oil pump. To prevent foreign bodies from being drawn into the pump, it passes through a gauze strainer at the lower end of the pick-up pipe. From the pump the oil is passed through a paper filter and then through a drilling in the cylinder block to the main oil gallery. The oil pressure in the filter and main gallery is about 400 kPa (4 bar or 60 psi). To ensure that the design pressure is not exceeded, and so prevent damage to the oil seals, an oil pressure relief valve is fitted in the main gallery. The oil, which is still at a pressure of about 400 kPa, goes through drillings in the block to the main bearings. The oil then passes through drillings in the crankshaft to the big-end bearings. At one end of the engine there is a long vertical drilling to take the oil supply to the camshaft and valve gear.

RACER NOTE

Oil pressure gauges typically read in psi, which may also be written lb/in^2 — during a race the oil pressure should normally remain constant within a given range — surges or drops outside of this range indicate possible faults.

Gear-type oil pump

The function of the oil pump is to draw oil from the sump and send it under pressure to the filter and on to the main gallery, from where it is distributed to other parts of the engine. The gear-type oil pump is used in a large number of vehicles. It is driven by the camshaft, usually by a skew gear mechanism. The skew gear drives only one of the two toothed gear wheels in the pump;

FIGURE 6.4
Hot hatches line up for the start at a club event

the other follows the driven gear. Oil enters the pump through the inlet from the pick-up pipe. Oil is carried round the pump in the spaces between the gear teeth; it is then forced out of the pump into the filter. The meshing of the gear teeth on their inside prevents the oil returning to the sump. If the teeth wear so that the clearance between the gear teeth and the casing exceeds 0.1 mm (0.004 in) the oil pressure will be reduced and the engine might become noisy, especially when it starts from cold when big-end bearing rattle might be heard.

Oil pressure relief valve

To prevent the engine oil from exceeding a preset figure, an oil pressure relief valve is located in the main oil gallery. Excess pressure may be generated if there a blockage in the gallery or the engine runs at high speed for a period. The oil pressure relief valve takes the form of a spring-loaded plunger, which is forced away from its seat when the preset pressure is reached. Excess oil from the gallery passes through the valve and returns directly to the sump. Removing oil from the main gallery reduces the oil pressure so that the pressure is reduced. When the pressure is lowered the valve is closed by the force of the spring, so the oil can no longer return to the sump.

Oil pressure warning light

The purpose of the oil pressure warning light on the dashboard is to warn the driver of low oil pressure. The dashboard light is activated by

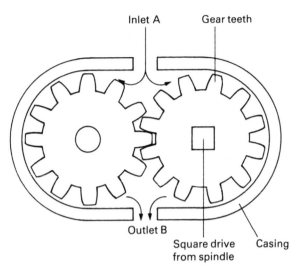

FIGURE 6.5
Gear type oil pump

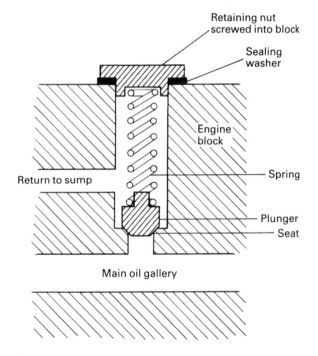

FIGURE 6.6
Oil pressure relief valve

a pressure-sensitive switch that is screwed into the main oil gallery. The switch makes the light glow when the oil pressure drops below about 30 kPa (5 psi). The light will also come on when the engine is switched on but is not running. Higher rated switches are available for tuned engines – typically 100 kPa (15 psi).

Oil filter

It is necessary to keep the engine oil clean, and free from particles of metal and carbon that could damage the inside faces of the oil pump or bearing surfaces. This is the job that the oil filter performs. All the oil passes through the filter before it goes to the bearings. The oil filter element is made from a special porous paper; this allows the oil to pass through it but holds back the foreign bodies. The oil filter element is housed in a metal canister that works as a sediment trap for the particles that the filter element has prevented from flowing into the main gallery.

Nomenclature

When the moving parts of the engine rub against each other, they wear. That is, tiny particles of metal are scraped off or dislodged into the sump or other parts of the engine. Also, the heat and combustion inside the engine creates particles of carbon or soot. These particles are referred to as foreign bodies. In filters, the heavier foreign bodies drop off the filter walls and fall to the bottom of the canister, or filter housing. This layer of trapping is referred to as sediment.

Servicing the engine lubrication system

Servicing the engine lubrication system involves changing the engine oil and replacing the oil filter. This is done at the intervals set down by the vehicle manufacturer. Typically this service interval is 16 000 km (10 000 miles). When carrying out lubrication system servicing the following points should be noted:

- The rubber seal on the oil filter must always be replaced.
- The sump plug's sealing washer must always be checked and replaced if necessary.
- Always check the oil level when the vehicle is on a level surface.
- After the engine has run, wait for about two minutes before rechecking the oil level and adding more oil.
- Never overfill nor under fill the engine; keep the oil level between the minimum (MIN) and the maximum (MAX) marks on the dipstick.

SAFETY NOTE

Beware of hot engine oil; never change the oil straight after the vehicle has completed a long or fast journey. The oil temperature can be over 110 degrees Celsius, and therefore there is a risk of scalding when removing the sump plug or filter. Always use barrier cream and protective rubber gloves when dealing with dirty engine oil to reduce the risk of contracting dermatitis.

Oil seals

To prevent oil leaking from rotating components an oil seal is needed. There are three different types of oil seals, namely:

- felt packing

- scroll seals
- lip seals.

Felt packing, as its name implies, is soft felt packed into a cavity. The rotating shaft turns in gentle contact with the felt. The felt becomes soaked in oil; this reduces the friction against the shaft. The oil also expands the felt so that the seal remains tight on the shaft and acts as an effective oil seal.

The scroll is a groove cut in the rotating shaft, which acts like an Archimedean screw. That is, as the shaft rotates, the screw draws the oil away from the open end of the shaft and returns it back to the sump.

The lip seal forms a hard, spring-loaded knife-edge against the rotating shaft. The pressure exerted by this knife-edge is very high and prevents any oil at all flowing past the seal, under even the most arduous conditions. The lip seal is retained in the oil seal housing by the interference fit of its outer rubber layer.

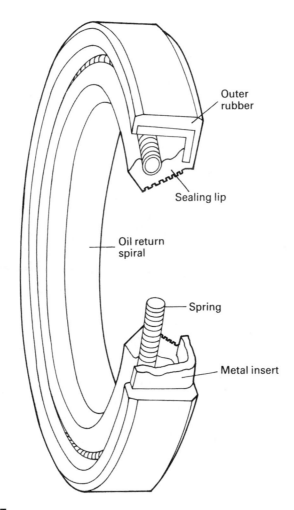

FIGURE 6.7
Lip type oil seal

95

6. TWO-STROKE PETROL ENGINE LUBRICATION

Lubrication of the two-stroke engine is by a petrol/oil mixture. The lubricating oil is added to the mixture of petrol and air so that it lubricates the crankshaft and big-end bearings when it is in the crankcase, while on its way to the combustion chamber.

The burnt lubricating oil is emitted from the exhaust in the form of blue smoke. Because of the effect that this has on the environment, two-stroke engines are only allowed in small motorcycles.

Petrol/oil mixture – low performance type two-stroke engines have their supply of lubricating oil added to the petrol when the tank is topped up. The oil lubricates the main and big-end bearings when the mixture is passing through the crankcase. The rider must add a quantity of oil to the petrol each time the motorcycle is refuelled. The amount of oil will depend on the ratio of petrol to oil that is required by the engine. A typical ratio is 40:1 (40 parts of petrol to 1 part of oil). In imperial units this works out at one gallon of petrol being mixed with one-fifth of a pint of oil. Often, the petrol filler cap is designed to incorporate a measure of the correct amount of oil to be added to one gallon of petrol. In SI units this will be one-eighth of a litre of oil to five litres of petrol. The oil used is special two-stroke oil that is made to mix easily with the petrol. However, it is normal practice to rock the motorcycle from side to side after the petrol and oil have been added to ensure that it has mixed well.

FIGURE 6.8
What you are most likely to see of a Ducati 1098

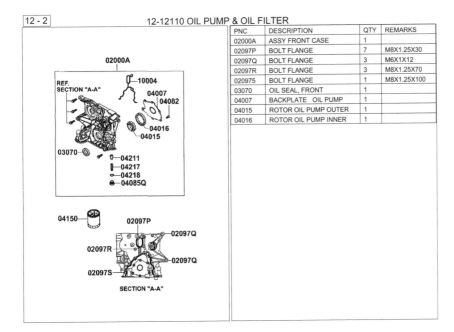

12-12110 OIL PUMP & OIL FILTER

PNC	DESCRIPTION	QTY	REMARKS
02000A	ASSY FRONT CASE	1	
02097P	BOLT FLANGE	7	M8X1.25X30
02097Q	BOLT FLANGE	3	M6X1X12
02097R	BOLT FLANGE	3	M8X1.25X70
020975	BOLT FLANGE	1	M8X1.25X100
03070	OIL SEAL, FRONT	1	
04007	BACKPLATE OIL PUMP	1	
04015	ROTOR OIL PUMP OUTER	1	
04016	ROTOR OIL PUMP INNER	1	

FIGURE 6.9
Oli pump and filter

Nomenclature

The term motorcycle is used in this book and unless they are otherwise especially distinguished, this term includes mopeds, scooters, go-peds and other two- or three-wheeled vehicles.

7. TOTAL LOSS SYSTEM

High performance two-stroke motorcycles have a separate oil tank, usually located under the rider's seat. Oil is pumped from this by a crankshaft-driven pump, so that it is mixed with the petrol at the carburettor. This ensures a constant mixture of petrol and oil in the correct proportions. It saves measuring out quantities of oil on the service station forecourt. Sometimes an adjustable regulator is fitted, so that the petrol/oil mixture can be adjusted to suit different engine running conditions.

MULTIPLE-CHOICE QUESTIONS

1. An oil labelled Hypoy 90 SAE is likely to be used in:
   (a) a car engine
   (b) a car gearbox
   (c) a bus axle
   (d) a car axle

2. The formula $\mu = F/W$ is used to find the:
   (a) amount of oil in the sump
   (b) oil viscosity

(c) coefficient of friction

(d) size of oil seals

3. The horizontal passage in the engine from which oil is fed to the main bearings is called the:

   (a) strainer

   (b) main gallery

   (c) relief valve

   (d) rocker shaft

4. The oil filter element is usually made from:

   (a) porous paper

   (b) wire wool

   (c) rubber

   (d) PVC

5. Felt packing, scroll and lip are all types of:

   (a) oil filter

   (b) oil seal

   (c) oil pump

   (d) two-stroke lubrication

6. To lubricate a two-stroke SI engine:

   (a) oil is mixed with the petrol

   (b) oil is added through the sparking plug hole

   (c) no oil is used

   (d) big-ends are pre-packed with grease

7. A lubricating oil that is made from crude oil pumped from the ground is called:

   (a) castor oil

   (b) engine oil

   (c) vegetable oil

   (d) mineral oil

8. Two functions of the lubrication system are to:

   (a) reduce friction and cool the bearings

   (b) reduce friction and improve combustion

   (c) warm the sump and prevent corrosion

   (d) increase friction and prevent corrosion

9. An oil with a high viscosity number is said to be:

   (a) thin

   (b) thick

   (c) fat

   (d) solid

10. The type of lubrication usually provided by grease is:

   (a) boundary

   (b) full-film

   (c) pressure

   (d) petrol/oil

(Answers on page 253.)

FURTHER STUDY

1. Using workshop manuals and manufacturers' information sheets, compile a chart to show the recommended lubricants for a few cars of your choice.
2. Dismantle an engine and follow the oil flow path from the sump, through the engine and back to the sump again.
3. Some components are lubricated by pressurized grease from a grease gun through a grease nipple. Describe how a grease gun works.

CHAPTER 7

Clutch

KEY POINTS

- The main components of the clutch are the pressure plate, the spinner plate and the thrust bearing.
- Most clutches have a diaphragm-type spring.
- The spinner plate may have a solid or a sprung centre.
- Clutch dust is a potential health hazard.
- Clutch adjustment is important to ensure that it does not slip.

The purpose of the clutch is to **transmit the torque**, or turning force, from the engine to the transmission. It is designed so that the drive can be engaged and disengaged smoothly and easily. By disengaging the drive, the clutch allows the gears to be changed smoothly, and it provides a temporary neutral position. This allows the transmission gears to be engaged or disengaged whilst the engine is running.

The clutch assembly is contained in the housing between the engine and the gearbox, on front-engined, rear-wheel drive cars this is called the **bell housing**; on other cars it is just called the clutch housing, or clutch casing. The main components of the clutch are the pressure plate, the spinner plate and the thrust bearing. The flywheel also has the job of being part of the clutch as well as its other two functions. The engagement and the disengagement of the clutch are carried out by a foot-pedal operated mechanism on most popular vehicles.

SAFETY NOTES

The dust from the clutch plate is a health hazard and breathing it in may cause respiratory problems — use a breathing mask if appropriate. Clutch components may also be hot to touch if the vehicle has just been running, so use of mechanics gloves is advised.

Basic Motorsport Engineering.

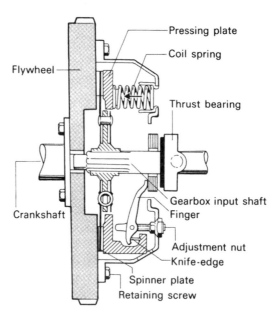

FIGURE 7.1
Coil spring clutch engaged

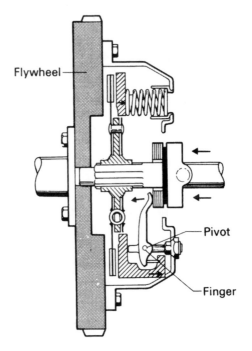

FIGURE 7.2
Coil spring clutch disengaged

1. TRANSMISSION OF TORQUE

The transmission of torque from the engine to the gearbox depends on the strength of the diaphragm spring in the pressure plate, the diameter of the spinner plate, the number of friction surfaces and the coefficient of friction of the clutch materials. The stronger the spring, and the larger the diameter of the spinner plate, the greater the torque which can be transmitted. The clutches of trucks are up to three times the diameter of those on cars, and they can weigh about ten times more. On motorsport vehicles the clutches have a small diameter, but have several friction surfaces, typically six plates, which means 12 friction surfaces.

Nomenclature

In this book the word truck is used to mean LGV, PCV, HGV and bus unless it is noted otherwise.

FIGURE 7.3
The author's kart racing team

2. DIAPHRAGM-SPRING CLUTCH

The diaphragm-spring clutch is used on most cars and light commercial vehicles. The diaphragm spring is shaped like a saucer, or a deep dinner plate, with a series of radially cut grooves.

In the engaged position, the diaphragm spring is shaped like a saucer. The force of the outer rim forces the pressure plate against the spinner plate. The pressure plate cover is bolted to the flywheel and the pressure plate is attached to the cover with flexible metal straps. When the flywheel rotates, the cover rotates; the cover turns the pressure plate by means of the **metal straps**. Most engines rotate in a clockwise direction when viewed from the front, as the straps must pull, not shove. This means that the clutch must be assembled so that the straps pull in an anti-clockwise direction when seen from the rear.

The diaphragm spring pivots on the cover using rivets with shoulders as **fulcrum points**. A fulcrum is another name for a pivot – something to swing on. It is against these rivets that the spring forces itself, to transmit force to the pressure plate and the flywheel to transmit the drive.

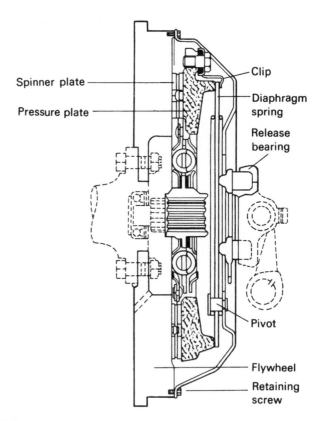

FIGURE 7.4
Diaphragm clutch engaged

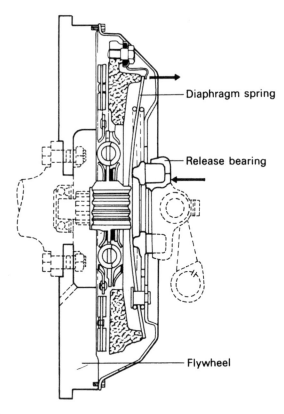

- Diaphragm spring
- Release bearing
- Flywheel

FIGURE 7.5
Diaphragm clutch disengaged

To disengage the clutch, the thrust race presses in the middle of the diaphragm spring. This causes the spring to pivot on the shoulders of the rivets so that it lifts its outer rim. This is similar to the action of a jam jar lid, or a CD in its case. The lifting of the outer rim pulls the pressure plate away from the spinner plate so that the spinner plate can rotate freely. The clutch is now disengaged so that the drive is not transmitted to the gearbox.

RACER NOTE

The diaphragm-spring clutch is used on most modern vehicles, but old race cars and other vehicles sometimes use coil-spring clutches. It is useful to compare the two as an example of the development of automotive technology and sound engineering principles. The diaphragm-spring clutch has the following advantages: it is smaller, having only one flat spring; it is lighter, as it uses less metal; it has fewer moving parts to break or wear. As it has only one spring it is consistently smooth.

3. SPINNER PLATE

The spinner plate consists of a **steel hub** which fits on to the splines of the gearbox input shaft. Attached to the hub is a disc that carries the friction material, which is riveted to the disc. The friction material on old vehicles is asbestos. Asbestos dust is a very serious health hazard; breathing it in can lead to asbestosis and subsequent death.

Modern vehicles use an asbestos substitute; however, you should remember that breathing in *any* dust is to be avoided as it can be harmful to your throat and lungs. Most workshops have a special vacuum cleaner to suck up clutch and brake dust. The clutch parts which are being reused should be cleaned with a special cleaner; usually this is called **brake cleaner** because it is used for cleaning brake dust (this is the same as clutch dust). The rags and other materials, along with the dust, must be disposed of safely in special plastic sacks.

There are two types of spinner plate in common use: the **solid centre** type and the **sprung centre type**. With the solid centre type the drive is directly from the friction lining to the hub. The sprung centre type has a series of coil springs to transmit the drive from the disc, which hold the friction disc to the hub. The springs serve two purposes:

- to absorb the shock loads when the clutch is suddenly engaged
- to absorb the small fluctuations in engine speed and vibrations, giving a smooth transmission of power to the gearbox.

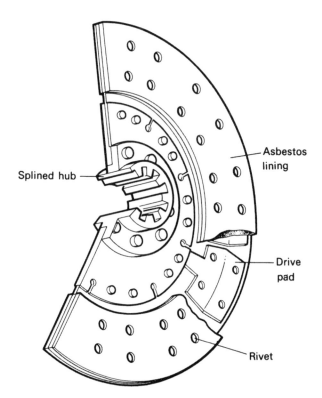

FIGURE 7.6
Solid centre spinner plate

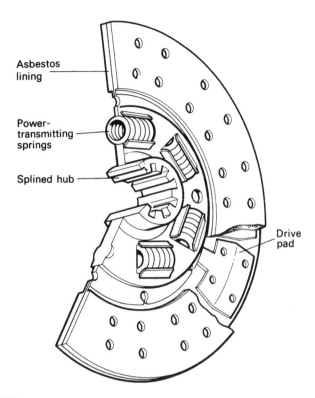

Asbestos lining

Power-transmitting springs

Splined hub

Drive pad

FIGURE 7.7
Sprung centre spinner plate

▌RACER NOTE

Motorsport vehicles usually have solid-centre clutch plates, often these are of the paddle type.

FAQs

What is meant by the clutch is engaged, or disengaged?

The clutch is engaged when it is in a position to transmit the drive from the engine to the gearbox; that is the clutch spring is holding the spinner plate tightly between the pressure plate and the flywheel, and the pedal is in the up position. When the driver presses the clutch pedal to the floor, the clutch is disengaged. This means the pressure is taken off the spinner plate so that it can rotate freely between the pressure plate and the flywheel. When the clutch is disengaged the drive from the engine is not transmitted to the gearbox.

4. OPERATING MECHANISMS

The clutch is disengaged by the driver pressing down the pedal on the left; it is engaged by releasing the pedal. When not operating the pedal, the driver must rest their foot away from the pedal. Some cars have footrests for this purpose.

The operating mechanism must have a small amount of free-play to ensure that the clutch is fully engaged. The pedal action operates the thrust race that moves the clutch diaphragm. The connection between the pedal and the thrust race may be by one of several different linkages — you'll find the popular ones are rod, cable and hydraulic; we'll look at each different type in detail.

Rod clutch

The method of operation for the rod clutch is the simplest. A steel rod is connected between the pedal and the clutch cross-shaft. When the driver depresses the **pedal,** the **rod** is pulled and the **cross-shaft** is rotated. The clutch cross-shaft transmits the movement from the rod to the clutch thrust bearing. You'll find these on old vehicles and some industrial plant such as forklift trucks. The problem with this mechanism is that it can transmit engine vibrations to the clutch pedal.

Cable clutch

The cable clutch is used on a large proportion of cars and vans. The inner, twisted wire **bowden cable** moves inside a steel outer guide cable (sometimes called a sheaf). The pedal pulls the inner cable, which in turn moves the clutch cross-shaft. The outer cable acts to provide a guide and controls the length for adjustment purposes. This system is similar to a bicycle brake or gear cable; it is adjusted in a similar way, that is by using a screwed nipple to change the length and take out the slack. The flexibility of the cable means that vibrations from the engine are unlikely to be transmitted to the clutch pedal.

Hydraulic clutch

Both the rod and the cable mechanisms rely on mechanical linkages; the lengths of the levers affect the mechanical advantage. A more sophisticated system is the hydraulic clutch. This system uses **hydraulic fluid**, like that used

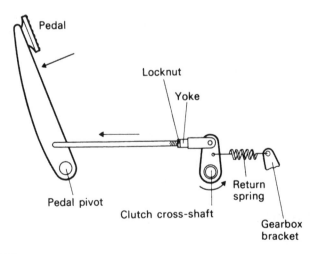

FIGURE 7.8
Rod clutch linkage

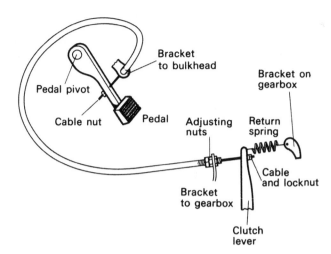

FIGURE 7.9
Cable clutch linkage

in hydraulic brakes, to transmit the movement from the clutch pedal to the cross-shaft in the bell housing. This system has its mechanical advantages; the difference between the force applied by the driver's foot on the pedal and the actual force that the clutch thrust race applies to the pressure plate diaphragm is built into the hydraulic system.

The clutch pedal moves a push rod, which in turn pushes a **hydraulic piston** into the clutch **master cylinder**. The hydraulic fluid above the piston is forced along the connecting tube. In turn the fluid forces the **clutch cylinder** piston against a short operating rod, which transmits the force to the clutch cross-shaft. These systems are very smooth in operation, and are used on many cars and trucks. It is essential to ensure that the clutch master cylinder reservoir is kept topped up to the correct level with the correct type and grade of hydraulic fluid. Most cars and trucks use the same fluid for both the brakes and the clutch.

Nomenclature
The mechanical advantage is the leverage gain, or torque multiplication, given by a mechanical mechanism. For instance, you might need to apply a force of 100 N to the clutch pedal and move the pedal 40 mm to disengage the clutch. The linkages in the clutch mechanism may have increased this force at the thrust race to 1000 N and reduced the distance travelled to 4 mm. The mechanical advantage would then be 10 to 1. It is the same principle as when you jack up a car weighing over a tonne using only one hand. The hydraulic jack uses the different sizes of piston to achieve the same results. The hydraulic clutch works in a similar way to the hydraulic jack.

5. CLUTCH ADJUSTMENT

It is essential that there is enough **free-play** in the clutch mechanism for the thrust bearing to be clear of the pressure plate. This is needed to ensure that the clutch does not slip. It is also important that there is not too much free-play, otherwise the clutch may not be able to be disengage completely.

109

The normal amount is 2 to 4 mm (one-eighth inch) of free-play at the thrust race, or cross-shaft lever. This will be the equivalent of about 20 or 25 mm (1 inch) of free-play at the pedal.

Adjustment is by screw threads on the cable of the operating arm. However, many clutches are self-adjusting by means of a ratchet type of device.

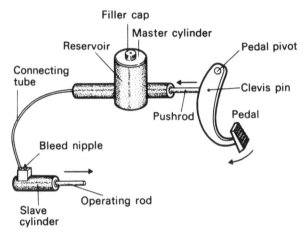

FIGURE 7.10
Hydraulic clutch linkage

FIGURE 7.11
Paul Presgraves — check him out at www.paulio-racing.com

21 - 4   21-25J05 CLUTCH MASTER CYLINDER & TUBES

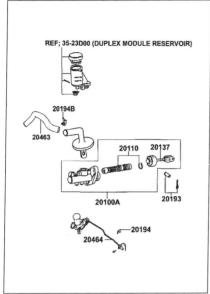

PNC	DESCRIPTION	QTY	REMARKS
20100A	CLUTCH MASTER CYL	1	
20110	PISTON KIT	1	
20137	PUSH ROD	1	
20193	PIN ASSY	1	
20194	CLIP HOSE	1	
20194 B	CLIP HOSE	1	
20463	HOSE RESERVOIR	1	
20464	CLUTCH, TUDE ASSY	1	

FIGURE 7.12
Clutch master cylinder and tubes

21- 2   21-21010 CLUTCH & CLUTCH RELEASE

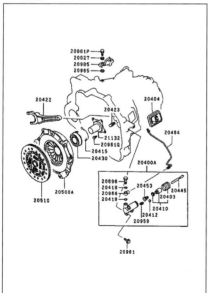

PNC	DESCRIPTION	QTY	REMARKS
20027	INSULATOR	1	
20400A	CYLINDER ASSY, CLUTCH RELEASE	1	
20403	BOOT, RELEASE CYLINDER	1	
20404	BOOT, RELEASE FORK	1	
20410	CYLINDER KIT, CLUTCH RELEASE	1	
20412	CAP, BLEEDER SCREW	1	
20415	CLIP, RETURN	1	
20418	GASKET	2	
20422	FORK ASSY, RELEASE	1	

FIGURE 7.13
Clutch master clutch release

6. CLUTCH FAULTS

The main faults likely to occur on a clutch are slipping and grabbing. Slipping is when the clutch is not transmitting the drive; this might be brought about by:

- the spring being fatigued or tired
- oil on the friction lining
- the friction lining being worn down to the rivets
- lack of free-play.

Grabbing is when the clutch cannot be engaged smoothly. This means the clutch suddenly grabs and takes the drive up with a thud. Grabbing may be caused by:

- spring damage or uneven wear
- wear in the mechanism or friction lining
- a broken spinner plate hub.

MULTIPLE-CHOICE QUESTIONS

1. Rod, cable and hydraulic are all types of:
   (a) clutch plate
   (b) clutch mechanism
   (c) bell housing
   (d) gearbox
2. The clutch which is the lightest and has the least moving parts is the:
   (a) multi-spring
   (b) coil spring
   (c) diaphragm spring
   (d) wet spring
3. Solid centre and spring centre are both types of:
   (a) diaphragm
   (b) pressure plate
   (c) spinner plate
   (d) release bearing
4. The clutch friction lining is made from:
   (a) steel
   (b) aluminium
   (c) carbon
   (d) an asbestos substitute
5. The amount of free-play at the cross-shaft should be about:
   (a) 1 mm
   (b) 4 mm
   (c) 14 mm
   (d) 40 mm
6. Clutch adjustment is needed to prevent:
   (a) clutch slip
   (b) clutch grab

(c) high pedal pressure

(d) low pedal pressure

7. The drive is transmitted from the pressure plate cover to the pressure plate by:

   (a) rivets

   (b) straps

   (c) bolts

   (d) nuts

8. The clutch spinner plate is connected to the gearbox input shaft by:

   (a) splines

   (b) slots

   (c) a bearing

   (d) a thread

9. Oil on the friction lining may cause:

   (a) clutch slip

   (b) clutch drag

   (c) clutch grab

   (d) a smooth take-off

10. As well as the pressure plate, spinner plate and thrust bearing, which of the following plays a part in the clutch operation?

    (a) distributor

    (b) starter motor

    (c) brake pedal

    (d) flywheel

(Answers on page 253.)

FURTHER STUDY

113

1. Look at a cable clutch and describe where the ends of the outer cable are attached.
2. Find out how clutches are operated on motorcycles.
3. There are other types of clutches in use other than those described in this chapter, carry out research to find the name of another type of clutch.

CHAPTER 8

Transmission System

KEY POINTS
- The transmission includes the gearbox, the final drive gears and the propeller shaft and/or the drive shafts.
- The gearbox allows the engine speed to be varied to suit the road conditions.
- The final drive gears give the vehicle its overall gear ratio.
- Universal joints and constant velocity joints are used in the transmission.
- To allow the driven wheels to turn at different speeds when cornering, a differential gear is used.
- Straight cut, helical, double helical and epicyclic gears are used.

1. COMPONENT LAYOUT

The gearbox is fitted behind the engine and the clutch on **conventional layout** vehicles — that is, with a front-mounted engine driving the rear wheels. The **bell housing** which covers the clutch is usually part of the **gearbox casing** and connects the gearbox to the engine. On conventional cars, a ring of bolts around the bell housing secures the gearbox to the engine. The weight of the gearbox is supported at the rear by a cross-member that attaches the gearbox to the chassis. The gearbox is held on the cross-member using a rubber mounting, this gives flexibility and prevents vibrations from being passed to the chassis and body.

A similar layout of components is used on both **front-wheel drive** (FWD) and **rear-wheel drive** (RWD) vehicles, except that the relative position of the gearbox in the vehicle is different. That is, the gearboxes of FWD engines are usually mounted across the engine compartment so that the gearbox is to one side. On some FWD cars the gearbox is below the engine. On RWD cars the gearbox may be to one side of the engine, or it may be mounted like a conventional layout but in reverse order. You should look at some typical

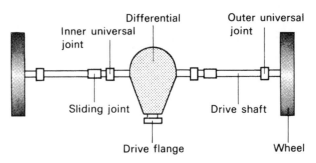

FIGURE 8.1
RWD layout

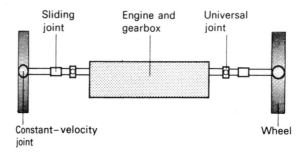

FIGURE 8.2
FWD layout

cars and trucks to be able to visualize the location of the gearbox and the other transmission components.

On **mid-engined** cars the engine is in front of the gearbox, they may be mounted **longitudinally** as on Formula Ford for example or **transversely** as on MG TF.

Propeller shaft

On conventional layout vehicles, which includes many vans and trucks, the gearbox is mounted to the rear of the engine and supported by the chassis; the rear axle is mounted on the road springs. The function of the propeller shaft is to transmit the drive from the rear of the gearbox to the rear axle, so propelling the car. The rear axle moves up and down with the road springs as the vehicle travels over bumps. This movement means that the **angle** of the propeller shaft between the gearbox and the rear axle changes as the car moves along the road. To accommodate changes in this angle, a moving joint is fitted at the ends of the propeller shaft. As the rear axle moves up and down, it also tends to rotate if it is mounted on leaf springs. The rear axle moves in an arc; this is called the **nose arc**. As the axle rotates, this arc causes

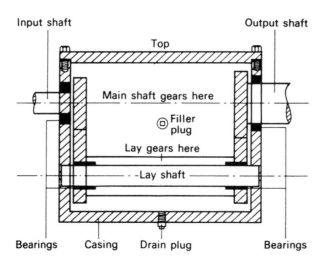

FIGURE 8.3
Gear box layout

the distance between the gearbox and the axle to change. To allow the propeller shaft to increase and decrease in length to accommodate movement of the axle, a **sliding joint** is fitted to the propeller shaft. The sliding joint allows up to 75 mm (3 inch) of variation in length of the propeller shaft.

The propeller shaft is a hollow metal tube; each end is welded to a flange to accommodate the universal joints and the sliding joint. A hollow tube is used because it is both light and strong.

117

█ RACER NOTE

If you change a gearbox and need a shorter propeller shaft, there are a number of engineering firms who will make up a custom propeller shaft and ensure that it is balanced correctly.

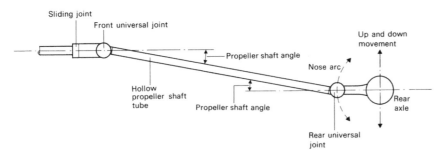

FIGURE 8.4
Propeller shaft layout

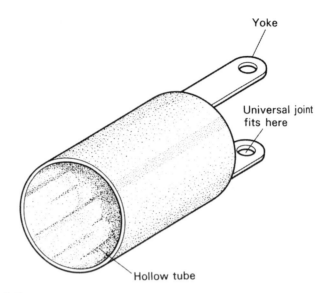

FIGURE 8.5
Propeller shaft section

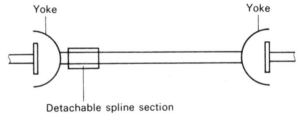

FIGURE 8.6
Propeller shaft yoke position

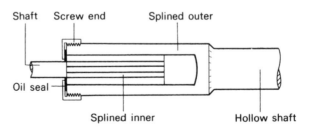

FIGURE 8.7
Propeller shaft splines

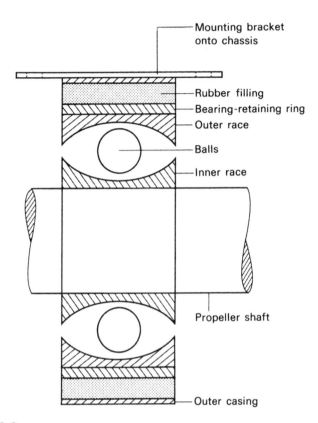

FIGURE 8.8
Propeller shaft centre bearing

119

Support bearing

On long vehicles it is necessary to use two short propeller shafts instead of one long one. This ensures that the power is transmitted from the gearbox to the rear axle smoothly, without propeller **shaft whip** or **wind-up**. The support bearing is a rubber-mounted bracket with a ball type bearing, in which the inner ends of the front and the rear propeller shafts rotate. The bracket is attached to the chassis between the gearbox and the rear axle. The support bearing is especially important at high speeds.

FAQs

What is the difference between whip and wind-up of the propeller shaft?
Whip is when the propeller shaft bends out in the middle. It forms a sort of bow shape like a skipping rope as it spins round. Wind-up is when the propeller shaft twists in torsion, and stores energy which can make the transmission jerk. The shape, if exaggerated, would look like a telephone handset cord, or a piece of twine.

Drive shafts

On rear-wheel drive (RWD), front-wheel drive (FWD) vehicles, and cars fitted with independent rear suspension (IRS), the drive shafts transmit the power from the differential to the rear driving wheels. The differential divides the drive between the two rear wheels, allowing the outer wheel to turn faster than the inner wheel when cornering. Drive shafts are like short propeller shafts. The drive shafts may contain a universal joint, a **constant velocity joint** (CV Joint) and a sliding joint. The constant velocity joint is needed to ensure that the drive is transmitted evenly at all times.

2. GEARBOX

Function

The job of the gearbox is to allow the car to accelerate and climb hills easily, and to provide a means of reversing. This is done by using a selection of gear trains that enable changes to be made in the ratio of engine speed to wheel speed, and in the case of reverse gear, the direction of rotation.

Principles of gearing

The reason for needing a gearbox is that the engine only develops usable power over a limited range of speeds − called the **power band**. The speed at which power is developed depends on the type of engine. For example, trucks develop their power at low speeds, typically 3000 rpm. Racing cars and motorcycles are the opposite, in that they develop their power at very high speeds; this can be up to 20 000 rpm. Most cars and vans develop their usable power between about 2000 and 5000 rpm. This means that if only one gear were fitted, in order to be able to set off from rest the car would need a gear ratio to give about 10 mph at 2000 rpm. At 5000 rpm the top speed would be 25 mph. At the other end of the scale, to have a top speed of 90 mph at 5000 rpm, 2000 rpm would give 36 mph. As you can see these are like first and top gears on a typical small car.

The gearbox acts like a lever, enabling a small engine to move a very heavy object. This is similar to how a tyre lever enables the tyre fitter to apply great force to the tyre bead.

Also remember that the gearbox provides a means of reversing the car, and a neutral gear position.

RACER NOTE

The gearbox has four main functions, it gives:

1. Low gears for acceleration, moving heavy loads and climbing steep gradients.
2. High gears to enable high-speed cruising.
3. A neutral gear, so that the engine can be running whilst the car is stationary.
4. A reverse gear so that the car can be manoeuvred into parking spaces and garages, for example.

Gear ratio

The gear ratio of any two meshing gears is found by the formula:

Gear ratio = Number of teeth on driven gear/
Number of teeth on driver gear
This is usually written = Driven/Driver

Where two gears mesh together, the gear ratio is :
Gear ratio = B/A
= 50/25
= 2/1
This is written 2:1 (say two to one).

This means that for each two turns of A, B will rotate once; hence two (turns) to one (turn). That is, gear B will rotate at half the speed of gear A. In other words, B will rotate at half the number of revolutions per minute compared with gear A.

If equal size pulleys and ropes were attached to the shafts to which gears A and B are fixed, as in Figure 8.10, it would be possible to use the 10 kg (22 lb) weight to balance the 20 kg (44 lb) weight. This is because the turning effort, or torque, is increased proportionally to the gear ratio. Although the speed is halved, the turning effort is doubled. This effect of the gear ratio is used when climbing steep hills or pulling heavy loads, such as a trailer.

3. GEARBOX RATIOS

The gears used in a typical gearbox are in **compound** sets of gears. For example, in first gear, four gear wheels are in mesh and transfer the power from the clutch to the propeller shaft. An example can be seen in Figure 8.12. With the input gear A and the two lay shaft gears B and C, and the output gear D. By using more gear wheels, in what is called a compound train, smaller gear wheels can give a bigger gear ratio whilst taking up less space.

To calculate the gear ratio in Figure 8.12, we have to decide which gears are driven, and which gears are drivers. As A is the gear that gives the input, this is a driver. Gear B is driven by gear A. gear C is attached to the lay shaft like gear B and therefore is turning at the same speed. So we can take C as a driver to D; D is therefore a driven gear.

RACER NOTE

Gear ratios are a bit confusing at first; it is a good idea to look at a sectioned gearbox to understand what is happening 'in the metal' as you might say. If you cannot get to see a sectioned gearbox, try to find an old gearbox that you can take apart. Also, most workshop manuals have lots of pictures of gearboxes, which might help you to understand how the gears run together.

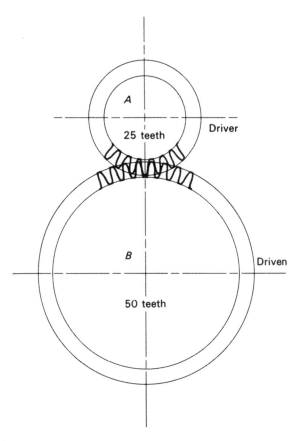

FIGURE 8.9
Gear ratio

Compound gear ratio = Driven/Driver × Driven/Driver
In our example :
  = B/A × D/C
If the number of teeth on each wheel is :
A = 10
B = 20
C = 15
D = 30
The formula would become :
Gear ratio = 20/10 × 30/15
  = 600/150
  = 4/1
  = 4:1

Final drive gear ratio

The final drive ratio is the ratio of the speed of the gearbox output to that of the road wheels.

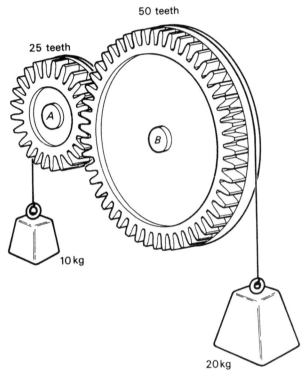

FIGURE 8.10
Turning effort

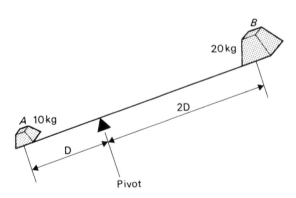

FIGURE 8.11
Gear leverage

The final gear ratio = Number of teeth on crown wheel
/Number of teeth on pinion
For example, with a 50 tooth crown wheel and a 10 tooth pinion:
Final drive ratio = 50/10
            = 5 : 1

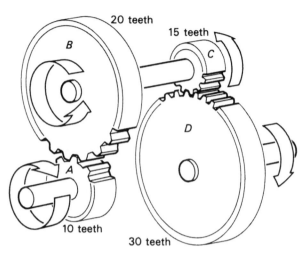

20 teeth
15 teeth
B
C
D
A
10 teeth
30 teeth

FIGURE 8.12
Compound gears

Overall gear ratio

The overall gear ratio is the ratio of the speed of the engine to the speed of the road wheels. This is found by multiplying the gearbox ratio by the final drive ratio.

$$\text{Overall gear ratio (OGR)} = \text{Gearbox ratio} \times \text{Final drive ratio}$$

If the gearbox ratio is 2.5:1 and the final drive ratio is 3:1, then:

$$\begin{aligned} \text{OGR} &= \text{Gearbox ratio} \times \text{Final drive ratio} \\ &= 2.5 \times 3 \\ &= 7.5:1 \end{aligned}$$

Layout

The power enters the gearbox through the input shaft that is splined at its outer end into the clutch spinner plate. The power is passed through the constant mesh gears to the lay shaft, then through the selected gear sets to the output shaft. The output shaft is connected to the propeller shaft that transmits the power, or turning force, onto the rear axle.

The gearbox casing is usually made from cast iron. It holds the gears firmly in place in relation to each other, and provides a reservoir for the lubricating oil.

Nomenclature

Many of the gearbox parts have alternative names, so if you are using a workshop manual you should always check out exactly which part is being referred to. Here are a few alternatives.

Input shaft – first motion shaft, primary shaft, clutch shaft, spigot shaft or jackshaft.

FIGURE 8.13
A motor club event in the Caribbean

Lay shaft – second motion shaft or counter shaft.
Output shaft – mainshaft, third motion shaft.

On FWD and RWD gearboxes the layout is the same as on conventional ones, except that the output shaft may directly connect to the final drive and differential gears instead of the propeller shaft.

4. GEAR TEETH

There are three main types of gear teeth used in gearboxes: spur gear, helical gear and double helical gear. Each kind of gear can be identified by the shape of its teeth.

Spur gear – this has straight teeth, like a cowboy's spur. This type of gear is also called straight cut. Straight cut gears can only carry a limited load and they are noisy in operation. You can hear straight cut gears occasionally rattle.

FIGURE 8.14
Spur gear

FIGURE 8.15
Helical cut gear

FIGURE 8.16
Double helical gear

Helical gear — so-called because if the shape of these teeth were projected, as around a long tube, the shape formed would be a helix. Another example of a helix is the screw thread on a bolt. Because the tooth is longer than the gear is wide, it is stronger than the equivalent straight cut gear. Helical gears are used in the gearboxes of most cars as they are quiet in operation. The disadvantage of helical gears is the side thrust. The two meshing gears will have a side thrust which increases with the applied load and the angle of the teeth. The lay shaft gears are machined as a unit, so the side thrust is passed through the metal. The main shaft gears are usually free to slide on the main shaft, so thrust washers are needed to hold them in place.

Double helical gears — are made like two rows of opposing helical gears. They are machined from one piece of metal so that the side thrust on one half of the gear balances the side thrust on the other half. This means there is no tendency for the gear to move sideways on the main shaft. Double helical gears are used in the gearboxes of trucks and buses where it is important to be able to transmit high loads.

5. EPICYCLIC GEARS

Epicyclic gears are used in automatic gearboxes and Sturmey Archer bicycle hub gears. This is a type of gear arrangement where sun and planet gears run inside an annulus. This arrangement allows a range of ratios to be obtained from one gear set. In Figure 8.17 the inner gear is called the sun gear (1), the outer toothed part is the annulus gear (2), and the small gears between the sun gear and the annulus are called planet gears (3). Various ratios can be obtained from one set of epicyclic gears, by locking each of the different sections in turn to the gearbox casing. For example, in an automatic gearbox the annulus can be held by a brake band so that the

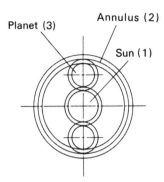

FIGURE 8.17
Epicyclic gears

power is transmitted from the sun gear to the planet gears. In this case the planet gears are turning, that is running around the inside of the annulus. If the carrier of the planet gears is held, the sun gear will rotate the planet gears on their spindles, which will turn the annulus. The latter gear ratio would be the lower.

This system of gearing is very compact when compared to conventional gearbox arrangements; this is the reason for its use in automatic gearboxes, cycle gear hubs and overdrive units. However, epicyclic gears are expensive to manufacture and they require great skill to assemble.

6. GEARBOX OPERATION

There are two types of manual gearboxes in use: the sliding mesh gearbox and the constant mesh gearbox. The sliding mesh gearbox is so-called because the gears slide into mesh with each other. The constant mesh gears get their name from being constantly in mesh with each other.

Sliding mesh gearbox

The general layout of the sliding mesh gearbox is shown in Figure 8.18. Power enters the input shaft from the clutch. This turns the input gear, which turns the lay shaft. The input gear and the lay shaft gear that turns with it are called constant mesh gears. The power is transmitted to the main shaft from the lay shaft by whichever main shaft gear is slid into contact. The engagement of each gear is detailed individually; the numbers are those of the gears in the sliding mesh gearbox diagram.

FIRST GEAR

First gear on the main shaft (8) is slid into mesh with the first gear on the lay shaft (14). So the power from the input gear (2) goes to the lay shaft (11), then from the first gear on the lay shaft (14) to first gear on the main shaft (8). The sliding gears are splined on the main shaft (9) so that the shaft turns when the gears are turned.

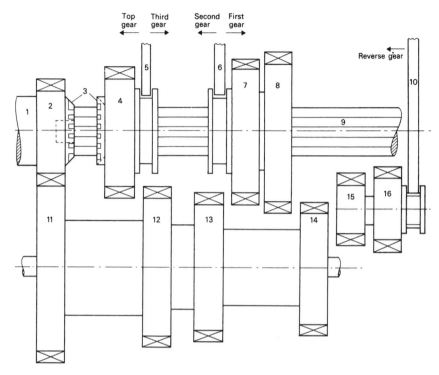

FIGURE 8.18

Sliding mesh gears

SECOND GEAR

Second gear on the main shaft (7) is usually connected to first gear (8) so that only one selector fork (6) is needed to engage first gear or second gear. The fork moves in backwards (right in Figure 8.18) for first gear, and forwards for second gear. The second gear on the main shaft (7) is slid along into mesh with second gear on the lay shaft (13) so that the power path is input gear (2) to lay shaft (11), second gear lay shaft (13) to second gear main shaft (7), through the splines to the main shaft (9).

THIRD GEAR

Third gear is engaged by meshing the third gear on the main shaft (4) with the third gear on the lay shaft (12). So that the power path is input gear (2) to lay shaft (11), third gear (12) to main shaft third gear (4) and that turns the main shaft (9).

FOURTH GEAR

Fourth gear is engaged by meshing the dog teeth (3) on the front of the third gear main shaft (4) with those on the input gear (2). The top gear is therefore direct drive. The input shaft turns the main shaft (9) at the same speed, the drive being through the dog teeth (3). The constant mesh gears (2 and 11) turn the lay shaft, but it is only idling, it is not transmitting any power.

REVERSE GEAR

Reverse gear is engaged by moving the reverse idler into mesh (15 and 16). The reverse idler is a shaft with two gears attached. One gear (16) meshes with the first gear on the main shaft, the other (15) with the first gear on the main shaft. The power path is thus input gear to lay gear to idler gear to first gear on the main shaft. Looking at the front of the gearbox the gears and shafts will rotate in the following directions:

Input gear: clockwise
Lay shaft gears: anti-clockwise
Reverse idler gear: clockwise
First gear main shaft: anti-clockwise

NEUTRAL

When no gears are in mesh this is neutral, and in this position no drive is transmitted.

FAQs

Does the lay shaft turn when the gearbox is in neutral?
Yes, the constant mesh gears are turning all the time, so the lay shaft must be turning too.

Constant mesh

All the gears, except reverse, are in constant mesh, not transmitting power, but idling, except when they are engaged. The main shaft gears are not splined to the main shaft, but run on bushes on the main shaft. This means that the gears and the shaft can turn independently of each other. Figure 8.19 shows the general layout. To engage gear, the synchromesh units that are splined onto the main shaft are slid into mesh with the dogteeth on the gears, so that the gear turns the synchromesh unit that transmits the power to the main shaft. The power paths of each gear are detailed in the following few paragraphs. The numbers refer to those on Figure 8.19.

Neutral

In neutral gear no power is transmitted as none of the synchromesh hubs are engaged with the gears. However, all the gears are turning when the engine is running and the clutch is engaged. The input gear turns the lay shaft through the constant mesh gears.

First gear

The synchromesh hub (9) is slid on the shaft (13) so that it engages with the dog teeth on the main shaft first gear (11). The power path is then input gear

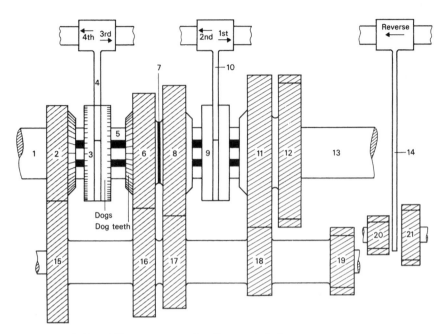

Gears 12, 19, 20 and 21 are spur gears, the others are helical gears

FIGURE 8.19
Constant mesh gears

(2) turns the lay shaft (15), which turns the first gear on the main shaft (11) through (18). The power then goes through the dog teeth to the synchro-mesh hub (9) and through its splines to turn the main shaft (13).

Second gear

The synchromesh unit is slid in the opposite direction to engage with the dog teeth on the side of the second gear on the main shaft (8). The power path is thus input gear (2) lay shaft (15 and 17), second gear main shaft (8), synchromesh unit (9), and main shaft (13).

Third gear

The other synchromesh unit (3) is slid into mesh with the third gear on the main shaft (6). The power path is similar to first and second gear, being input gear (2), lay shaft (15 and 18), third gear main shaft (6), dog clutch (3) and the main shaft (13).

Fourth gear

Fourth gear is engaged by sliding the synchromesh unit (3) into mesh with the dog teeth on the side of the input gear (2). This gives direct drive in the same way as the sliding mesh gearbox. That is through the dog teeth, the synchromesh hub and the splines to the main shaft.

Reverse gear

The reverse gear is usually engaged in the same way as in the sliding mesh gearbox. That is the idler gears (20 and 21) are slid into mesh with a reverse gear on the lay shaft (19) and one on the main shaft (12).

Nomenclature

The sliding mesh gearbox is also referred to as a crash gearbox because of the noise made by inexperienced drivers using them. The constant mesh gearbox is often called the synchromesh gearbox because of its use of synchromesh hub units. Ferdinand Porsche invented the synchromesh hub unit when he worked on VW cars in the 1930s. You will not find a fully sliding mesh gearbox on a modern ordinary car, but you will find cars which have one or two gears of the sliding type and the rest synchromesh. You will also find sliding mesh gears on some racing cars and some motorcycles.

You should also note that the first motion shaft gear and the lay shaft gear which it meshes with are called constant mesh gears too.

7. SYNCHROMESH HUB

A synchromesh hub is shown in Figure 8.20. It consists of a hub, which is splined to the main shaft, and an outer sleeve, which is splined to the hub. Spring-loaded balls hold the outer sleeve in the neutral, or the engaged, position.

When initial pressure is applied by the selector to the outer sleeve, the pressure of the ball also moves the hub section. This causes the conical surface on the hub to give the initial interference to the cone on the gear involved. This interference adjusts the speeds of the mating shafts to the same speed. That is, it synchronizes them. Further pressure by the selector pushes the outer sleeve over the balls against the spring pressure until they engage in the next row of grooves. At this point the outer hub has engaged its inner splines with the dog teeth on the gear. Therefore the gear is fully

131

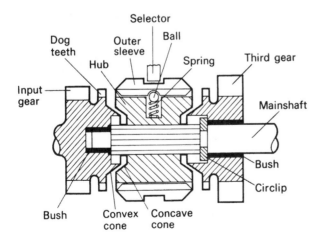

FIGURE 8.20
Synchromesh hub

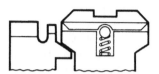
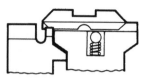

Synchromesh unit operation: initial interference of cones' synchronizing speeds

Synchromesh unit operation: drive gear engaged; outer sleeve over dog teeth

FIGURE 8.21
Synchromesh operation

engaged. Drive from the gear is passed from the dog teeth to the outer sleeve, to the hub and to the shaft.

Selector

Figure 8.22 shows a typical selector, or selector fork to give it its full name. The fork is moved by a selector rod through a mechanism attached to the gear lever.

Detents

To ensure that the gears are held in the selected position, a system of detents is used. Examples of detents are shown in the Figures 8.23, 8.24

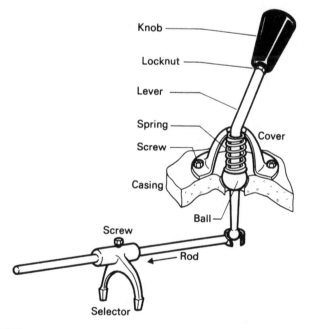

FIGURE 8.22
Gear lever

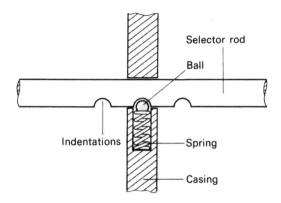

FIGURE 8.23
Selector detent

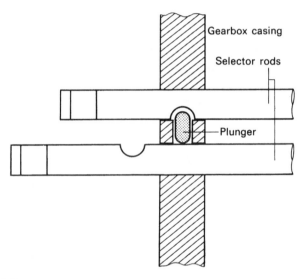

FIGURE 8.24
Selector detent

and 8.25. One uses a spring-loaded ball, another uses a plunger. The C-shaped detent is a mechanical locking device.

8. GEARBOX REMOVAL

You will usually need to remove the gearbox to replace the clutch, as well as to repair the gearbox itself. Most gearboxes are located underneath the vehicle; this means that you will need a hoist, or a pit, to gain sufficient access. On most cars the gearbox is connected to the engine by a ring of bolts around the bell housing; some form of rubber mounting to the body or chassis will also be used. The exact procedure for removing a gearbox will be

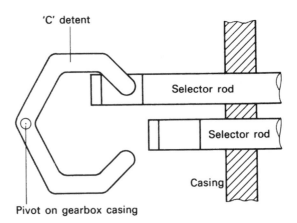

FIGURE 8.25
C-type selector detent

given in the workshop manual appropriate to the vehicle, but the general procedure is as follows:

- Disconnect the battery earth terminal for safety.
- If possible drain the gearbox oil.
- Remove the speedometer cable.
- Disconnect the gear lever mechanism.
- Disconnect the clutch mechanism.
- Remove the drive shafts or the propeller shaft.
- Support the engine.
- Use a cradle to support the gearbox.
- Remove the bell housing bolts and the mounting bracket.
- Withdraw the gearbox on the cradle.

RACER NOTE

Gearboxes are very heavy; you should always work with a colleague to get a team lift when removing them.

9. GEARBOX STRIPPING

The method of stripping gearboxes varies with the make and model of the vehicle. You should check with the workshop manual before carrying out any stripping to prevent damage to the gearbox or personal injury.

10. GEARBOX LUBRICATION

The gearbox oil varies in type and viscosity with the vehicle, so you must check the workshop manual before topping up or changing it. It is common to find that the same oil is used in the gearbox and in the engine, such as a 10/40 SAE. However, many vehicles, especially trucks, use oil such as EP 90 SAE.

11. FINAL DRIVE ARRANGEMENTS

The rear axle

Trucks and a few other conventional layout vehicles have live rear axles. That is, the axle transmits the driving force as well as supporting the vehicle. A typical rear axle carries inside it the following components:

- final drive gears
- differential gear assembly
- half-shafts
- rear hubs and bearings
- lubricating oil.

Final drive gears

The final drive may be in the form of a crown wheel and pinion in a truck, or other conventional layout vehicle rear axle; or an output pinion and differential gear in a FWD vehicle. On conventional layout vehicles, the crown wheel and pinion turn the drive through 90 degrees as well as providing an additional gear ratio. On FWD, RWD and mid-engine cars with a transverse layout, there is no need to turn the drive through an angle.

There are a number of different types of bevel gears used for the final drive gears; three of them are straight bevel gears, spiral bevel gears and hypoid bevel gears.

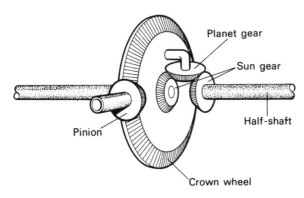

FIGURE 8.26
Final drive gears

Nomenclature

Bevel gears are ones cut so that they mesh to form a 90 degree angle. That is, they turn the drive through a 90 degree angle.

Straight bevel gears — these have straight cut teeth, so they are very noisy in operation.

Spiral bevel gears — have the pinion on the same centre line as the crown wheel, like the straight bevel gears, but the teeth are cut at an angle so that they are both strong and quiet in operation.

Hypoid bevel gears — these gears have the centre line of the pinion below that of the crown wheel. This is to allow the propeller shaft to be as low as possible, so the vehicle floor can be low and the handling is more stable. The gear teeth are both cut at an angle and curved. Hypoid bevel gears must be lubricated with special hypoy oil to withstand the very high pressures that are caused by these two gears rolling into mesh.

Worm and wheel — is an alternative to the more usual crown wheel and pinion. The worm looks like a very large bolt; it turns the gear teeth on the outside of the wheel. The wheel looks similar to a crown wheel, but the teeth are on the outer perimeter, not at an angle. The worm may be located below the wheel to give low chassis height, or above the wheel to give maximum ground clearance. The advantage of the worm and wheel drive is that it can transmit very high loads and give very low gear ratios. The gear ratio is calculated by dividing the number of teeth on the wheel by the number of starts on the worm. The term starts means the number of threads that go to make up the worm; this is usually between three and five.

$$\text{Gear ratio} = \text{Number of teeth on wheel}/\text{Number of starts}$$

> ## RACER NOTE
>
> The worm and wheel is a very important engineering principle. You will only find worm and wheel drive on specialized vehicles, such as heavy-duty dump trucks and off-road contractors' plant. However, you will find the principle applied in a number of ways in engineering; a good example is the screw driven hose clip, the most popular one being known by the trade name Jubilee Clip.

12. DIFFERENTIAL

The differential gears are mounted inside a carrier; the **crown wheel** is screwed to the outside of the **carrier**. The differential carrier is mounted on bearings that are carried by the axle casing. The differential gear components comprise two sun gear wheels and two planet gear wheels. The sun gear wheels are attached one to each end of the axle shafts; and the planet gears are free to rotate on a cross-shaft, which is located in the differential carrier.

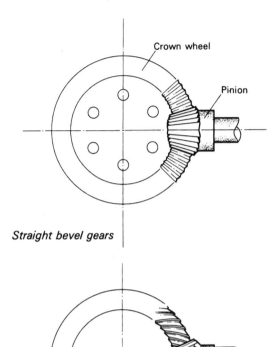

Straight bevel gears

Spiral bevel gears

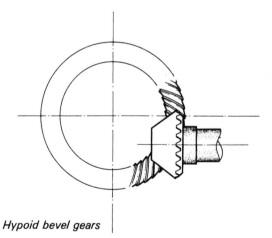

Hypoid bevel gears

FIGURE 8.27
Crown wheel and pinion set-ups

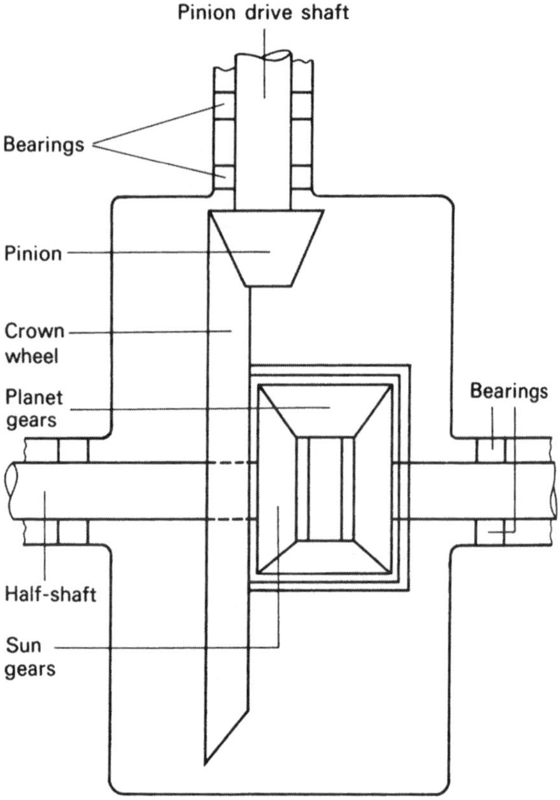

FIGURE 8.28
Differential layout

The functions of the differential are to:

- allow one wheel to turn faster than the other does. For instance, when going round a corner, the outer wheel will travel further and hence faster than the inner wheel
- divide the driving force, or torque, evenly between both wheels.

Operation of the differential

Figure 8.28 shows the essential gears — it does not represent a complete differential. Under normal straight-line conditions the pinion rotates the crown wheel which turns the differential carrier. The carrier turns bodily, taking the planet gears with it. It is important to note that the planet wheels do not spin on the cross-shaft; they pass the driving torque onto the sun wheels, so that the sun wheels turn at the same speed as the crown wheel, and both axles are therefore rotated. The speed and the turning force are the same for both axles and therefore both of the road wheels.

On cornering, the inner wheel drive shaft and the sun gear on the end of it are slowed in comparison to the outer one. Therefore there is a speed differential (difference). The planet wheels rotate around the slowed sun gear, spinning on their cross-shaft. This movement is passed onto the outer sun gear, which therefore turns faster. The outer sun gear gains the speed lost by the inner sun gear.

Half-shafts

The half-shaft connects the sun gear wheel to the hub, so transmitting the drive to the road wheel. The inner end of the half-shaft is splined to the sun gear wheel in the differential. The outer end of the half-shaft has a taper or a flange assembly to connect to the hub.

13. HUB/AXLE ASSEMBLIES

The rear hub and axle assembly is subjected to considerable forces, namely:

- the driving force to the wheels
- the braking force
- the load of the vehicle.

These forces put bending and shear stresses on the rear axle. To provide sufficient strength without undue weight, one of three types of hub/axle bearing arrangements may be used. The three arrangements are semi-floating, three-quarter floating and fully floating. Figures 8.29, 8.30 and 8.31 illustrate the different arrangements.

Semi-floating

The semi-floating assembly is the weakest arrangement because the bearing is between the inside of the axle casing and the half-shaft. The problem is, if it breaks the wheel will fall off completely.

139

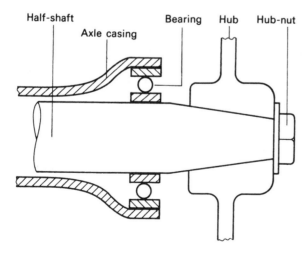

FIGURE 8.29
Semi-floating hub arrangement

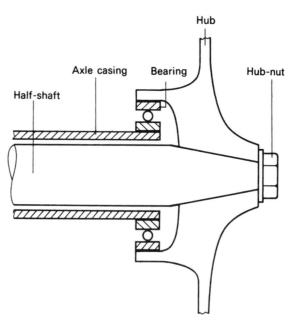

FIGURE 8.30
Three-quarter floating-hub arrangement

Three-quarter floating

The three-quarter floating arrangement is much stronger than the semi-floating type. The bearing is located between the hub and the axle casing. The major part of the weight is supported by the bearing – hence its name.

Fully floating

The half-shaft in this arrangement carries none of the vehicle's weight, hence the name fully floating. The hub is supported by twin ball, or roller, bearings. This arrangement is used on the axles of almost all trucks and buses. An added advantage is that both the half-shaft and the differential can be changed without jacking up the vehicle.

14. DRIVE SHAFTS

Drive shafts connect the differential gears to the hubs, to allow for movement they are usually fitted with some form of joint.

Hooke joints

The Hooke joint was invented by the physicist Robert Hooke (b 1635 – d 1703). Although the Hooke joint is its proper name, it is known more commonly as the universal joint.

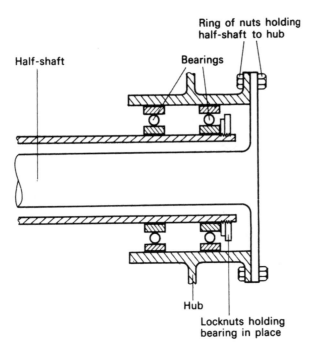

FIGURE 8.31
Full-floating hub arrangement

Nomenclature

The universal joint is another name for the Hooke joint. In the motor industry it is also known as a Hardy Spicer joint; Hardy Spicer is the trade name of the main manufacturer. This is similar to calling a vacuum cleaner a 'Hoover'.

The Hardy Spicer joint is constructed from a cruciform member that rotates in four cups, each containing needle roller bearings. Two of the cups fit into

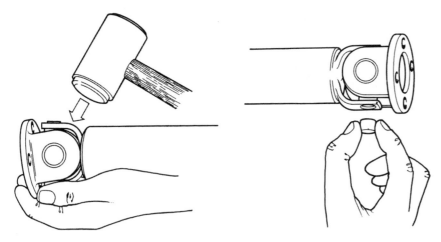

FIGURE 8.32
Removing a U/J

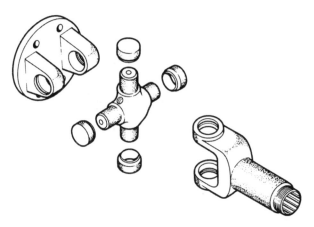

FIGURE 8.33
Hardy Spicer joint

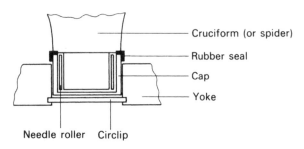

FIGURE 8.34
Needle roller bearings

the flange that is attached to the rear of the gearbox — the opposite two are attached to the flange at the end of the propeller shaft. At the other end of the propeller shaft is a similar arrangement for the final drive pinion. The needle roller bearings are usually pre-packed with special grease so that they are sealed for their life. On some vehicles the joints are lubricated by the means of grease nipples. A grease gun must be used to apply grease every 15 000 km (10 000 miles).

Hardy Spicer joints are fitted in pairs to give an even transmission of power. The yokes must be fitted to the propeller shaft in the same plane. Where a detachable sliding joint is fitted, it is possible to refit it with the yokes out of alignment; this will lead to an uneven transmission of power and make the vehicle shake.

CV joints

To give a drive without fluctuations in speed a constant velocity (CV) joint is used. This is important on front-wheel drive cars, where using a Hooke joint would lead to steering vibrations. The joint can transmit drive

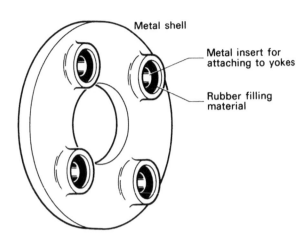

FIGURE 8.35
Layrub coupling

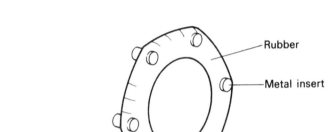

FIGURE 8.36
Rotaflex coupling

through an angle of more than 15 degrees without variations in velocity.
The CV joint is usually fitted in the centre of the suspension upright.
A popular type of joint is the Birfield joint; this uses an inner driving
member with grooves and an outer driven member, also with grooves.
There are balls in a cage that run in both grooves so that the power is
transmitted from the inner part of the joint to the outer part through the
balls. The balls are simply large ball bearings. The CV joints transmit all the
driving forces so to ensure that they are correctly lubricated a special grease
is used.

REPLACING A DRIVE SHAFT JOINT GAITER

The CV joint must be kept correctly lubricated and free from water and dirt.
As well as the practical requirements of protecting the CV joint, or the sliding
joint, it is a requirement of the MOT test that the drive shaft gaiters are kept in
good condition.

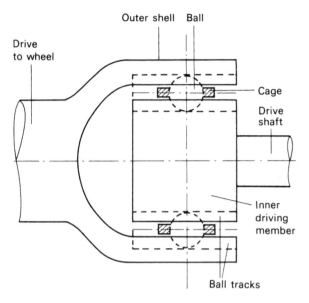

Ball driving joint (the Birfield constant velocity joint has curved tracks)

FIGURE 8.37
Constant velocity joint (CVJ)

To replace a gaiter on a CV joint it is necessary to remove the road wheel, disconnect one of the suspension joints and swing the suspension upright to one side, then pull back the damaged gaiter so that the CV joint can be freed from the drive shaft.

Before you fit a new gaiter you should always ensure that the joint is clean and free from water, and then apply the special CV joint grease before fitting the new gaiter.

RACER NOTE

Transmission components are, because of where they are on the car, likely to be very dirty; and the oil or grease that is used collects dirt easily. Where possible, you should always try to clean off the components before you work on them. There are many types of degreasers. If you use one which does not need water then there will be less risk of the part rusting. However, when you are working with any degreaser you must use protective gloves and goggles and be careful to keep it off your skin.

15. CHECKING TRANSMISSION OIL LEVEL

Most gearbox oil levels are checked using a **level plug**. That is, the gearbox has a plug on its side; oil is added through the hole when the plug is removed. To add the oil you need a squeeze bottle and a plastic

FIGURE 8.38
Vyrus — can you see how to sleer it?

22 -16   22-22130 M/T GEARSHIFT CONTROL

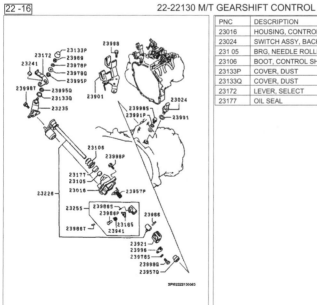

PNC	DESCRIPTION	QTY	REMARKS
23016	HOUSING, CONTROL	1	
23024	SWITCH ASSY, BACK LIGHT	1	
231 05	BRG, NEEDLE ROLLER	1	20
23106	BOOT, CONTROL SHAFT	1	
23133P	COVER, DUST	1	
23133Q	COVER, DUST	1	
23172	LEVER, SELECT	1	
23177	OIL SEAL	1	

FIGURE 8.39
Gearbox shift control

tube. The oil level is correct when the oil is level with the lower part of the hole.

Some gearboxes have short **dipsticks** attached to a bung on the top of the gearbox. You need to replace the plug to check the oil level.

In all cases, remember to only check the oil level when the vehicle is on a **level surface**. Also, be aware that gearbox oil can be hot, so let it settle for a short while after driving the car.

MULTIPLE-CHOICE QUESTIONS

1. The component which allows one wheel to turn faster than the other whilst cornering is the:
   (a) differential
   (b) final drive
   (c) hub
   (d) half-shaft
2. The formula for finding a gear ratio is:
   (a) driver/driver
   (b) driven/driver
   (c) driven/driven
   (d) driver/driven
3. If a vehicle has a long propeller shaft it may be fitted with a:
   (a) Birfield joint
   (b) third universal joint
   (c) support bearing
   (d) flange end
4. Lay shaft, main shaft and input shaft are all found in the:
   (a) gearbox
   (b) rear axle
   (c) differential
   (d) rear hub
5. Sun gear, planet gear and cross-shaft are all found in:
   (a) epicyclic gear trains
   (b) automatic gear boxes
   (c) synchromesh hubs
   (d) differentials
6. The input shaft of the gearbox is splined to the:
   (a) main shaft
   (b) lay shaft
   (c) synchromesh hub
   (d) clutch spinner plate
7. The gaiters on the CV joint must be checked and replaced if necessary to:
   (a) protect the CV joint from damage due to loss of grease
   (b) improve the gear ratio
   (c) enable better acceleration
   (d) make it easier to turn the steering

8. The type of hub and axle assembly used on trucks and buses is:
   (a) three-quarter floating
   (b) semi-floating
   (c) epicyclic
   (d) full floating
9. Which of the following types of gearbox uses helical gears:
   (a) sliding mesh
   (b) automatic
   (c) synchromesh
   (d) racing car
10. The gearbox is usually attached to the engine by the means of:
   (a) cir-clips
   (b) a ring of bolts
   (c) splined shafts
   (d) rubber mountings

(Answers on page 253.)

FURTHER STUDY

1. Pull apart an old gearbox and using the workshop manual identify the parts and calculate the gear ratios.
2. Using workshop manuals, or data sheets, compare the gear ratios for a number of different vehicles of your choice. If possible include some trucks.
3. Using an old drive shaft, or propeller shaft, remove and dismantle the CV joint or universal joint.

CHAPTER 9

Suspension and Steering

KEY POINTS

- The suspension and steering are combined at the front of the vehicle. Rear suspension is usually much simpler.
- Three popular types of suspension systems are beam axle, MacPherson strut and wishbone.
- Three popular types of springs are coil springs, leaf springs and torsion bars.
- Suspension joints must be checked at major services and the MOT test.
- Castor, camber and KPI are important angles with regard to handling.
- The Ackermann angle affects true rolling motion of the wheels.
- Wheel alignment is checked using optical gauges.

149

The front suspension and the steering mechanism are combined into one unit that carries out two distinct functions. The rear suspension is usually separate from other functions. On front-wheel drive cars, the drive shafts with their CV joints pass through, and turn with, the steering and suspension. In this chapter we are only looking at the suspension and the steering

1. FUNCTION OF THE SUSPENSION MECHANISM

The function of the suspension is to **connect the road wheels to the body/chassis**; it is designed to prevent the bumps caused by road-surface irregularities from reaching the occupants of the car. This is to make the car both pleasant to ride in and easy to drive. In the case of goods vehicles, the suspension protects the load from damage. You could consider how we would transport eggs if trucks did not have good suspension. The suspension also protects the mechanical components from road vibrations, so making them last much longer.

The suspension consists of a system of movable **linkages**, and a **spring** and a **damper** (also called a **shock absorber**) for each wheel. The **tyre** also forms part of the suspension; the flexing of the tyre absorbs small irregularities in the road and helps to keep the noise to a minimum. You might imagine what it was like before cars were fitted with pneumatic rubber tyres.

2. FUNCTION OF THE STEERING MECHANISM

The steering mechanism is needed to guide the car along its chosen path. The principle of steering is that the **front wheels exert a force to guide the car in the chosen direction**; the rear wheels will follow the front ones. The front wheels must exert enough force on the road to ensure that it maintains its set course. Steering the front wheels, that is turning them from side to side, is achieved by a set of **rods** and **levers** which are operated by the steering wheel acting through either a steering box or a rack and pinion unit.

Before we look at the actual components involved in the suspension and the steering mechanism we will define some of the special terms which are used. For example **castor**, **camber**, **kingpin inclination**, **Ackermann principle** and **wheel alignment**.

Castor

Castor, or to give its full name, the castor angle, is the angle that the **swivel pin** (or kingpin on trucks) is arranged to lean backwards in the **longitudinal plane**. By giving the swivel pins a castor angle of one or two degrees, the imaginary centre line meets the road before the centre of the wheel. This distance is called **castor trail**. The castor angle gives the front wheel a self-aligning, or self-centring, action. The wheel follows the pivot, so keeping the car on a straight course and reducing the need to manually straighten the steering wheel after negotiating a corner. This can be likened to the castors on a trolley. The castors swing round so that the trolley can be pushed in a straight line. You can see this in action on any supermarket trolley.

> **RACER NOTE**
>
> The angles shown for castor, camber and KPI in the following figures are greatly exaggerated so that you can see them easily. The actual angles are very small, generally between 0.5 and 2 degrees. To measure them you need specialised equipment.

Camber

Camber is the inclination of the road wheel in the **transverse plane**. The wheel is generally inclined outwards at the top; this is called **positive**

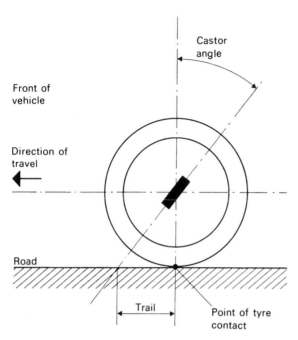

FIGURE 9.1
Castor angle

camber. If the wheel were inclined inwards at the top, it would be **negative camber**.

Nomenclature

The kingpin is a metal bar, or pin, that truck suspension pivots on. This type of suspension used to be used on cars too. The term kingpin is also used for any kind of suspension pivot mechanism, especially when referring to the suspension angles. If a car does not have a kingpin it can still have a kingpin inclination angle; that is the angle that the suspension has that is the equivalent of a kingpin on an older car.

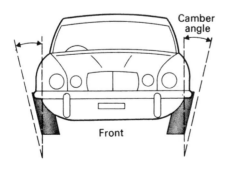

FIGURE 9.2
Camber angle

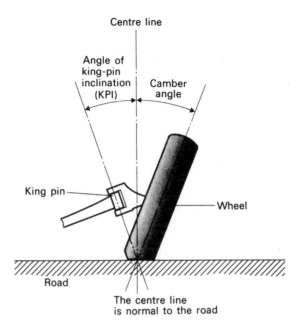

FIGURE 9.3
KPI and camber

Kingpin inclination (KPI)

KPI is the inclination of the kingpin in the **transverse plane**, which is the same plane as the camber. Normally, kingpin inclination is inwards at the top; this is **positive** KPI. **Negative** KPI is when the kingpin is inclined outwards at the top.

When the camber and KPI angles meet at the road surface this is called **centre-point steering**. The actual intersection of the lines is in the middle of the patch where the tyre meets the road.

Ackermann angle

When a vehicle turns a corner, all its wheels must rotate about a **common point**, or the tyre treads will be **scrubbed**. If you think of the vehicle going round in a full circle, all the wheels must be turning about the centre point of that circle. This is achieved by taking the radius lines from the centres of the wheels and steering them so that they are tangential to the radius lines. That is, the position of the wheels and the radius lines are at 90 degrees. The two rear wheels are on the same radius line; this is because they cannot be steered. The front wheels can be steered as they are on separate axes. The steering mechanism is designed so that the **inner wheel is always turned through a greater angle than the outer**. In Figure 9.5 A and B are the angles through which the wheels have to be turned. The inner angle A is greater than the outer angle B.

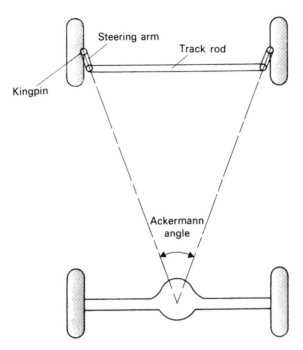

FIGURE 9.4
Ackermann angle

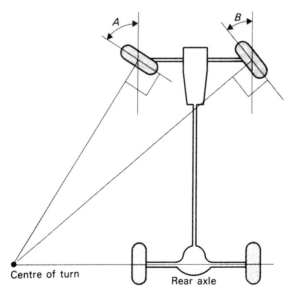

FIGURE 9.5
Ackermann principle

FAQs

What is true rolling motion?
True rolling motion is when the wheels roll without side-slip. If the car is travelling in a straight line, and the wheel alignment is set at zero, that is, there is no toe-in or toe-out, then the wheels will be rolling truly. If the car, with Ackermann, is turning a corner slowly, then all the wheels should roll truly.

Wheel alignment (toe-in and toe-out)

The front wheels are arranged so that they either **toe-in** or **toe-out**. Toe-in is when the front of the wheels is closer than the rear of the wheels. That is, A is less than B in Figure 9.6. Toe-out is when A is a greater than B. Typically toe-in or toe-out is 2 mm. You should look in the **workshop manual**, or on the **data chart**, to find out the setting for any particular vehicle. If the wheel alignment is not set correctly the tread will be scrubbed off the front tyres. That is because the wheels are not rolling truly, the wheel and the tyre move sideways; this will cause the tread to be scrubbed, like using an eraser on a pencil drawing.

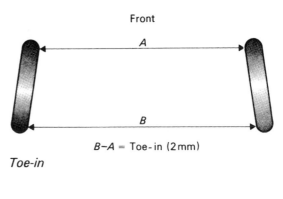

Front

A

B

$B-A$ = Toe-in (2mm)

Toe-in

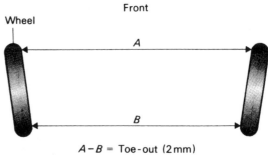

Front

Wheel

A

B

$A-B$ = Toe-out (2mm)

Toe-out

FIGURE 9.6
Wheel alignment

FIGURE 9.7
Adjustable suspension on a 750 Motor Club car

3. SPRINGS

The spring absorbs the shock when the wheel hits a bump in the road. The spring must be strong enough to support the vehicle and its load, but be able to be compressed when a bump is hit. There are a number of different springs: the popular ones are coil springs, leaf springs and torsion bar springs. If you look at the following figures you'll be able to see the differences. All these different springs are made from specially tempered, medium carbon steel called spring steel.

Coil spring

The coil spring is usually fitted around the shock absorber or MacPherson strut. It is made from round-section spring steel that is wound to shape. Coil springs offer a large amount of suspension movement and lightness.

Leaf spring

These are made from a number of flat metal sections, rather like thin leaves. The main leaf has an eye at each end so that it can be attached to the chassis. Leaf springs can carry a lot of weight, but suspension travel is limited.

Torsion bar spring

The torsion bar is a round bar which twists when it is loaded. It has the advantage that its weight is carried fully on the chassis; it can also be fitted in positions where height is limited. The disadvantage is that the amount of suspension travel that is allowed is very small.

4. SUSPENSION LAYOUT

There are lots of different suspension layouts; this section looks at some of the more popular ones.

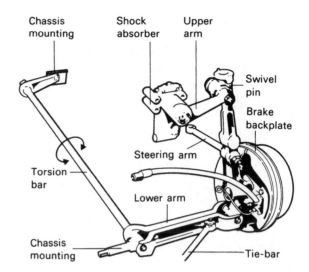

FIGURE 9.8
Torsion bar suspension

Beam axle suspension

The axle beam goes transversely across the vehicle; it is attached to the chassis with leaf springs. The leaf spring consists of a main leaf which has an eye at each end and several other supporting leaves. Both the axle beam and the spring are made from medium carbon steel. The front eye of the spring is attached to the chassis with a fixed shackle; the rear eye has a swinging shackle. The spring is attached to the axle with U-bolts. The steering mechanism is achieved with a kingpin and reverse Elliot linkage.

The disadvantage of beam axle suspension is that when one of the wheels hits a bump, the wheel on the other side is also tipped. This is because both wheels are attached to the same component that is the axle beam. As you can

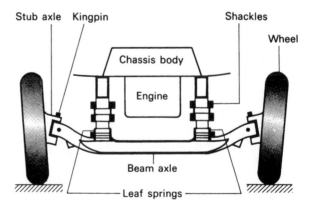

FIGURE 9.9
Vehicle suspension component

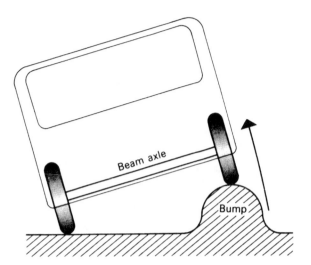

FIGURE 9.10
Beam axle hits bump, whole car is moved

see in Figure 9.10, when one wheel hits a bump the whole vehicle is made to tip up sideways.

Independent suspension

With independent suspension, each wheel is suspended independently of the others. Some vehicles have independent suspension on the front wheels only – this is called **independent front suspension (IFS)**. At the rear it is called **independent rear suspension (IRS).** When one wheel of a car with independent suspension hits a bump, only that wheel is deflected upward; the other wheels are not affected and the car remains level.

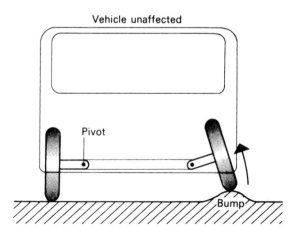

FIGURE 9.11
Independent suspension hits bump, only suspension components move

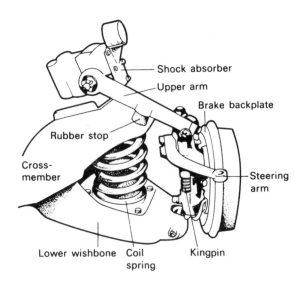

Shock absorber

Upper arm

Brake backplate

Rubber stop

Cross-member

Steering arm

Lower wishbone Coil spring

Kingpin

FIGURE 9.12
Wishbone suspension

Wishbone suspension

Wishbone suspension is so-called because the arms are shaped like chicken wishbones. The coil spring is usually mounted **concentrically** over the shock absorber. The shock absorber is attached to the lower wishbone and the chassis.

MacPherson strut suspension

The MacPherson (pronounced mac'fur'son) strut is a combination of a suspension swivel pin and a shock absorber unit in one assembly. The spring is a coil spring, which is mounted concentrically on the MacPherson strut. MacPherson strut suspension is very easy to remove from the vehicle: the top mounting goes to the inner wing and the lower one to the track control arm. With the strut removed you can then replace the faulty part while working on the bench. However, if you are replacing any of the components which require removal of the spring, then you need to compress the spring a little to prevent it flying and possibly causing damage or personal injury. Compressing the spring involves fitting spring compressors (which are sort of hooks with screw threaded parts) to two of the coils and screwing up the threaded parts until they take the spring load off the mountings.

5. SHOCK ABSORBERS

The purpose of the shock absorber is to dampen the spring action and reaction. The shock absorber stops the vehicle from bouncing each time it hits a bump in the road. There are two types of shock absorbers, these are

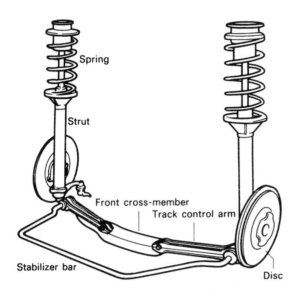

Spring

Strut

Front cross-member

Track control arm

Stabilizer bar

Disc

FIGURE 9.13
MacPherson strut suspension

telescopic and lever arm and they are easily identified by their shape and mountings.

FAQs

Is a damper the same as a shock absorber?
Yes.

Telescopic shock absorber

The telescopic shock absorber gets its name from its telescope-like shape and action. The cylinder is filled with a special type of oil, called shock absorber fluid. The lower part of the cylinder is connected to the suspension with a mounting eye. The upper part of the shock absorber, which comprises the piston and valve assembly, is attached to the vehicle's chassis with the upper mounting eye. The piston and the valve assembly move up and down inside the cylinder with the movement of the suspension.

When the wheel hits a bump, the suspension travels upwards, shortening the distance between the mountings. In this situation the piston travels down the cylinder. The resistance of the fluid slows the movement of the piston, so dampening the shock load on the suspension. When the wheel has travelled over the road bump, the suspension rebounds. That is, the wheel travels down again and the piston moves up in the cylinder as the distance increases again. The resistance of the fluid dampens the suspension movement and prevents the suspension from travelling too far. The shock absorber therefore

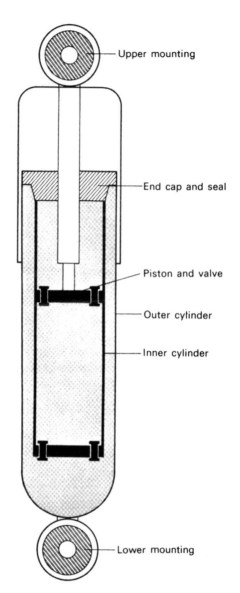

Upper mounting

End cap and seal

Piston and valve

Outer cylinder

Inner cylinder

Lower mounting

FIGURE 9.14
Telescopic shock absorber

dampens the suspension movement and stops the vehicle from bouncing like a rubber ball every time it hits a bump.

Nomenclature

Bump is when the suspension is compressed; that is the wheel goes up into the wheel arch and the body goes down towards the road. Rebound is the opposite of bump; that is the wheel goes down and the body goes up. Commentators on rallies often use the word 'jounce', this is another word for bump — you'll see it when the car lands back on the road after going over a humpback bridge. This take-off and jounce is referred to as 'yumping', the Scandinavian pronunciation of jumping.

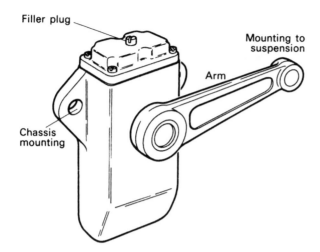

FIGURE 9.15
Lever arm shock absorber

Lever arm shock absorber

The lever arm shock absorber works in a similar way to the telescopic shock absorber. The body of the shock absorber is bolted to the vehicle's chassis; the lever arm is attached to the suspension and therefore moves up and down with the suspension. The arm is attached to a rocker assembly inside the body. The rocker assembly moves two pistons in parallel bores. One piston goes up as the other piston goes down, this is called double acting. The piston movement is given resistance by shock absorber fluid in the same way as the telescopic one. Lever arm shock absorbers usually have provision for their fluid to be topped up — telescopic ones generally do not.

161

Bounce test

When you are checking the shock absorbers on a vehicle you should:

- inspect the mounting for damage or wear — the rubber bushes should be tight
- inspect the body for damage or dents
- check the seals for fluid leaks
- carry out a bump test — that is press down on the wing and let it go quickly. It should not go up and down more than three times.

6. STEERING LINKAGES

Various layouts of steering linkages are used, depending on the type of vehicle and the system chosen by the designer. Figure 9.16 shows the layout of a typical truck steering system. Cars tend to use rack and pinion steering mechanisms; these are shown later in this chapter. The steering wheel is

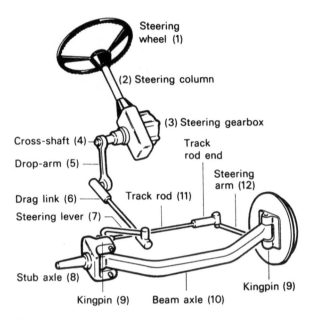

FIGURE 9.16
Steering layout

situated inside the car, this is mounted on the steering column that passes
through the bulkhead and connects to the other components that are
underneath the front of the car. The job of the steering linkages is to
convert the movement of the steering wheel into movement at the road
wheels.

When the steering wheel is turned, this turns the steering column that
operates the mechanism of the steering box. The cross-shaft is the output
from the steering box, which moves the drop-arm. This pushes or pulls the
drag link, which operates the steering lever, which is attached to the offside
stub axle. The offside wheel is thereby moved in the required direction. The
wheel hub is mounted on the stub axle, which pivots on the kingpin in the
beam axle. The track rod is attached to the offside steering lever so that it
moves transversely when the offside wheel is turned. This moves the nearside
track rod end (TRE), which is attached to the nearside steering arm, which
steers the nearside stub axle and the nearside wheel.

Steering box

The steering box is bolted to the chassis. The steering column is attached to the
steering box; the steering wheel is attached to the inner part of the steering
column. The cross-shaft is attached to the steering linkage underneath the car.
The steering box does a number of different jobs, the main ones are:

- Turning the drive through a right-angle (90 degrees) between the steering
  column and the cross-shaft.

- Giving a reduction gear ratio of about 14:1 so that the turning force applied by the driver to the steering wheel is increased at the cross-shaft. The movement of the cross-shaft will be reduced by an equivalent (14:1) movement ratio.
- Stops bumps caused by road surface irregularities being passed on to the driver.

Figure 9.17 shows a worm and peg steering box. The peg is at the end of the rocker shaft. The tip of the peg sits in the worm. The worm is fitted on the end of the steering column. When the steering column is turned by the steering wheel, the worm turns too. The peg follows the helical thread of the worm, this moves the rocker shaft which is attached to the cross-shaft. The cross-shaft turns moves the steering linkage.

It is important that the steering box is kept properly lubricated; usually a gearbox type of oil such as SAE 80 EP is used. The oil is added through the filler plug, which is also a level plug. That is, oil is added until it reaches the lower part of the filler hole. Some steering boxes have a separate grease nipple for the end bearing. The steering box can be adjusted to compensate for wear. There is an adjusting screw for wear in the peg and the worm. The bearing adjustment is by shims under the end plate.

The normal service interval for checking the steering box oil level and the adjustment is 15 000 km (10 000 miles).

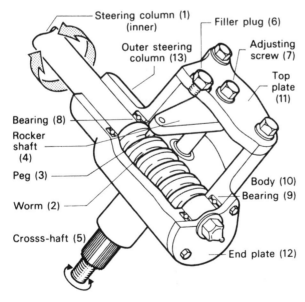

FIGURE 9.17
Steering box

RACER NOTE

To check for the correct adjustment to the peg and the worm follow this simple procedure:

- Jack the front of the car up and support it on axle stands.
- Set the steering wheel to the straight-ahead position. Move the steering wheel about 90 degrees in each direction; you should feel a slight tightness in the middle (straight-ahead) position.
- If this cannot be felt screw the adjusting screw down until it can, taking note that some adjusting screws have locknuts.

To check the end bearing this is what you do:

- Sit inside the car and see if the steering wheel will move vertically on the steering column.
- If there is excessive vertical lift then remove shims from between the end plate and the steering box body.

Rack and pinion

The rack and pinion assembly does the same job as the steering box and track rod combined together. The rack and pinion is a long and thin tubular

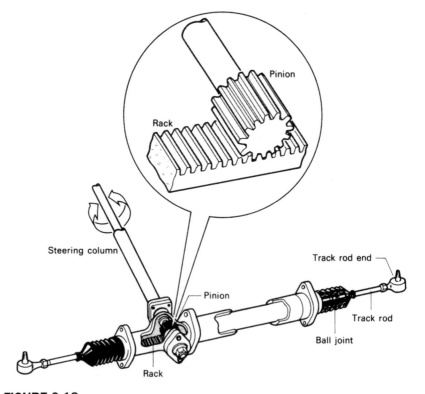

FIGURE 9.18
Rack and pinion steering

looking arrangement. It is used on most cars because it has the advantages of being both light and cheap.

The rack and pinion is made in one unit, which is bolted to the bulkhead with U-bolts. The steering column connects to the pinion. The track rod ends (TRE) are screwed onto the track rods. The track rods are attached to the rack by ball joints. The TREs connect to the steering arms on the hub carriers using taper fit pins.

The pinion is attached to the lower end of the steering column so that it is rotated when the driver turns the steering wheel. The pinion meshes with the rack, so that turning the pinion moves the rack to one side or the other. When the rack moves the track rods, steering arms and then the road wheels move.

7. WHEEL BEARINGS

Car hubs usually run on pairs of ball bearings. Trucks usually use roller bearings. The bearings may be either preloaded, or they may be adjustable.

Checking wheel bearings

With the car supported on axle stands, there are two checks to be made. One is for bearing free-play, the other for noise. To check for free-play, hold the top of the wheel with one hand and the bottom with your other hand. Try to move the wheel from side to side. If it moves more than the smallest amount, then the bearing needs either adjusting or replacing. To check for noise, spin the wheel by hand and listen; a good bearing should spin freely and quietly.

Checking steering and suspension joints

Two of the main points of wear in the steering system are the ball joints and the track rod ends. To check these joints for wear, the two parts to which the ball joints connect should be pulled in a direction in which you would expect them to part. Using hand pressure, maybe with a small lever, a joint in good condition should show no signs of free-play at all.

Adjusting wheel alignment

The wheel alignment, toe-in or toe-out, should be checked every 20 000 miles (30 000 km). This is carried out using optical gauges. The optical gauges may use a laser beam and may have a digital read out, depending on the level and sophistication of the technology.

The wheel alignment optical gauges are fitted to the front wheel, the laser beam is focused and the reading can then be taken. The reading scale may be in mm (inches on older gauges), or in degrees. Most wheel alignment gauges are supplied with a chart, or table, so that degrees can

FIGURE 9.19
Checking the corner weights on a car which used to belong to 'The Stig'

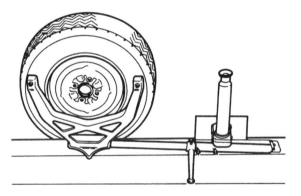

FIGURE 9.20
Wheel alignment gauge

be converted into mm for most standard size wheels; that is 10 inch to 20 inch.

8. RALLY CAR SPECIFICS

Most rally cars are based on road cars — the changes to the suspension depends on the specific competition regulations. Typically the changes will involve the springs and the dampers. The springs are chosen to give the suspension setting needed for the type of rallying — typically they are stiffer. The dampers are likely to be adjustable to allow variations in suspension behaviour — firmer, or harder, dampers are used to reduce suspension bounce. Lots of rallies are run off-road, using special stages, or white roads, so the suspension has to be robust and able to deal with hitting bumps at high speeds. To prevent damage additional bump or rebound stops may be fitted.

9. RACE CAR SPECIFICS

Unlike rally cars, race cars usually run on smooth circuits; however, the circuits tend to be tighter and the bends are taken at very high speeds. So, the amount of suspension travel is likely to be very small — as low as 3 mm (one-eighth of an inch).

The suspension movement will, however, be very rapid, and dampers with additional cooling will be required as well as fine adjustment for stiffness. The suspension on open wheel cars is very light and may be made of carbon fibre composites. The springs are very stiff and little compression is likely.

Most single-seat race cars use inboard suspension so that the weight is taken on the chassis.

Race cars tend to use negative camber to keep the tyre treads on the road, and very little castor, so that the car is not easy to keep in a straight line, but will corner without effort.

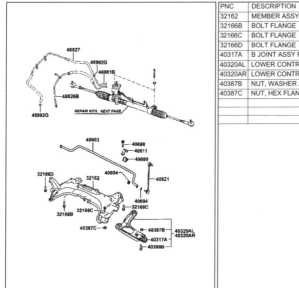

33 - 2		33-21A00 FR,CROSSMEMBER MODULE		

PNC	DESCRIPTION	QTY	REMARKS
32162	MEMBER ASSY	1	
32166B	BOLT FLANGE	2	M12X60
32166C	BOLT FLANGE	4	M12X80
32166D	BOLT FLANGE	2	M12X80
40317A	B JOINT ASSY FR	1	
40320AL	LOWER CONTROL ARM COMPL LH	1	
40320AR	LOWER CONTROL ARM COMPL RH	1	
40387B	NUT, WASHER ASSEMBELED	6	10
40387C	NUT, HEX FLANGE	2	12

FIGURE 9.21
Cross member module (suspension and steering)

167

| 33 - 6 | | 33-21A24 MACPHERSON STRUTS & SPRINGS | | |

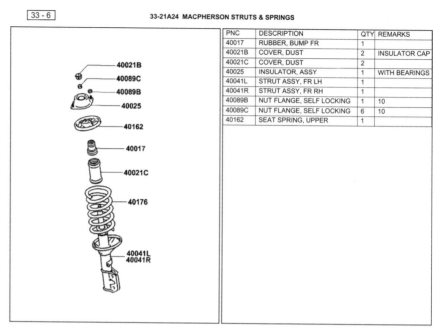

PNC	DESCRIPTION	QTY	REMARKS
40017	RUBBER, BUMP FR	1	
40021B	COVER, DUST	2	INSULATOR CAP
40021C	COVER, DUST	2	
40025	INSULATOR, ASSY	1	WITH BEARINGS
40041L	STRUT ASSY, FR LH	1	
40041R	STRUT ASSY, FR RH	1	
40089B	NUT FLANGE, SELF LOCKING	1	10
40089C	NUT FLANGE, SELF LOCKING	6	10
40162	SEAT SPRING, UPPER	1	

FIGURE 9.22

MacPherson struts and springs

| 34 - 5 | | 34-21B26 REAR STABILIZER | | |

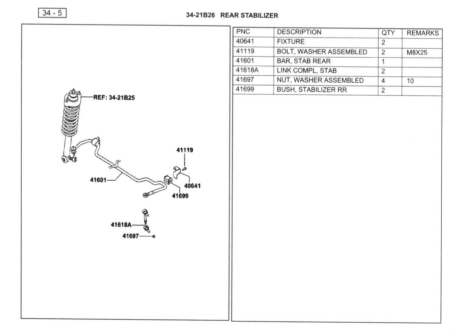

PNC	DESCRIPTION	QTY	REMARKS
40641	FIXTURE	2	
41119	BOLT, WASHER ASSEMBLED	2	M8X25
41601	BAR, STAB REAR	1	
41618A	LINK COMPL, STAB	2	
41697	NUT, WASHER ASSEMBLED	4	10
41699	BUSH, STABILIZER RR	2	

FIGURE 9.23

Rear stabilizer

MULTIPLE-CHOICE QUESTIONS

1. Beam axle, MacPherson strut, and wishbone are all types of:
   (a) suspension systems
   (b) steering systems
   (c) springs
   (d) shock absorbers

2. Even tyre wear, through true rolling motion, is achieved by the means of:
   (a) the camber angle
   (b) the Ackermann angle
   (c) rack and pinion steering
   (d) setting the toe-in

3. The component which is located between the track rod and the steering arm is the:
   (a) steering rack
   (b) steering box
   (c) track rod end
   (d) steering column

4. Telescopic and lever arm are both types of:
   (a) steering box
   (b) wheel bearings
   (c) steering angle
   (d) shock absorber

5. Hubs are lubricated with:
   (a) wax
   (b) water
   (c) grease
   (d) oil

6. Self-centring action is achieved by means of the:
   (a) camber angle
   (b) castor angle
   (c) kingpin inclination
   (d) toe-in

7. Which of the following is not a type of independent suspension?
   (a) MacPherson strut
   (b) beam axle
   (c) double wishbone
   (d) torsion bar

8. When the centre lines of the KPI and the camber intersect at the road surface, this is known as:
   (a) centre-point steering
   (b) Ackermann
   (c) castor
   (d) turning circle centre

9. The upper mounting of a telescopic shock absorber is usually attached to the:
   (a) wing

169

      (b) chassis
      (c) wishbone
      (d) axle beam
10. Worm, peg and drop-arm are all parts of the:
      (a) steering rack
      (b) steering box
      (c) MacPherson strut
      (d) beam axle

(Answers on page 253.)

FURTHER STUDY

1. Using the diagram of the Ackermann angle as a guide (Figure 9.4), make a model of the steering linkages. Turn the inner wheel in steps of 5 degrees, and measure how many degrees the outer wheel turns. If possible try this on a car using turntables.
2. Draw up a table of the steering angles for a range of different vehicles and see if there is a particular pattern to them.
3. Obtain either an old steering rack or steering box and strip it down using the workshop manual — can you name all the parts?

CHAPTER 10

Wheels and Tyres

KEY POINTS

- The wheels and the tyres must be perfectly round, rigid and correctly balanced.
- There are rules relating to wheel and tyre fitment and maintenance which must be followed, for example the minimum tread depth is 1.6 mm.
- Wheels may be made from steel or aluminium alloy; with either a well base or detachable flanges for fitting the tyre.
- The tyre and the wheel diameter are measured where the tyre fits the rim; the width is measured between the flanges.
- Different types of tyre construction and tread designs are used for different purposes.
- Specialist machines are used for changing tyres and balancing the wheel and tyre.

1. FUNCTIONS OF WHEELS AND TYRES

The wheels and tyres have a number of jobs to perform, which are to:

- allow the vehicle to freely roll along the road
- support the weight of the vehicle
- act as a first step part of the suspension
- transmit to the road surface the driving, the braking and the steering forces.

2. REQUIREMENTS OF WHEELS AND TYRES

For the wheels and the tyres to be able to carry out their functions efficiently they must be made and maintained to the following basic requirements:

1. They must be perfectly round so that they roll smoothly.
2. They must be balanced so that the steering does not shake.
3. They must be stiff to give responsive steering and smooth running.

3. WHEELS

There are many different types of wheels in use, we'll look at the most common ones, that is **steel well based**, **aluminium alloy**, **wire-spoked** and **two-piece**.

SAFETY NOTES

- Always use axle stands, with the vehicle on a flat and level surface when removing wheels.
- Never exceed the maximum tyre pressure given by the manufacturer.
- Always use a torque wrench to correctly tighten the wheel nuts.

Basic construction and sizes

Basically all wheels comprise of a **rim** and a **wheel centre**, which are attached together in some way. The rim is the part to which the tyre is fitted.

The rim has a **flange** to hold the tyre in place, a seating part to seal the tyre bead against and retain the air and a well section so that the tyre can be fitted and removed from the rim.

The wheel centre is the part that is attached to the hub. The wheel centre usually has four holes for the **wheel studs**. The back of the wheel centre has a flat section to make contact with the hub.

The **wheel diameter**, which is also the tyre size, is measured at the tyre seating part of the rim. You should note that the flange extends beyond this part of the wheel. The **wheel width**, which is also the equivalent of the nominal tyre width, is measured between the inside faces of the flanges.

FIGURE 10.1
Tyre van at a club event

Steel wheels

The rim and the wheel centre of the steel wheel are both made from pressed low-carbon steel. The rim is spot welded to the wheel centre. Steel wheels have the advantages of being cheap, tough and reasonably light in weight. Steel wheels are ideal for off-road events such as 4×4 cross-country and rallycross, as they are tough so will withstand a limited amount of impact damage.

Aluminium alloy wheels

Aluminium alloy wheels were originally developed for aircraft; they give a combination of extreme lightness and high stiffness. The aluminium is alloyed with silicone for wear resistance, copper for hardness and magnesium for easy casting. Aluminium alloy wheels are usually cast in one piece, but some specialist wheels are made in two parts that are held together with a ring of bolts. This construction allows the rim to split into two parts for easy fitting of the tyre. Two-piece alloy rims are flat across the whole section – that is they do not have a well section as it is not needed.

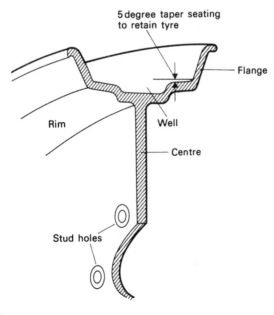

FIGURE 10.2
Well-based wheel

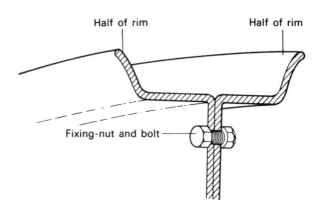

FIGURE 10.3
Two-piece wheel

The disadvantages of alloy wheels, as they are referred to, are that they are expensive, brittle and have a limited life span. The brittleness is a problem if the wheel hits a kerb or other hard object (called kerbing); in this case the rim is likely to chip or crack, which can cause the tyre to deflate.

Alloy wheels can be made in a variety of styles, so they are available with looks to suit the car and its owner. However, good looks are in the eye of the beholder.

Wire-spoke

The wire-spoked wheel is only used on a small number of sports cars. Unlike the other wheel types, which are attached to the hub with studs, the wire-spoked wheel is attached to the hub with a splined section and a single **rudge nut**. The wire-spoked wheel is slightly flexible and springy. Spoked wheels, as they are usually abbreviated, have the advantages of being light, good looking and allow cool air to pass through them to cool the brakes.

The disadvantages are that they are easily buckled, especially if the spokes become loose, and that they need an inner tube to seal in the air.

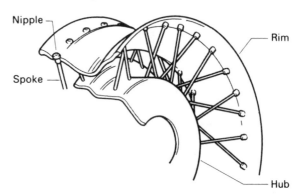

FIGURE 10.4
Wire-spoked wheel

4. TYRES

Construction

The basic construction of all vehicle tyres is very similar. The main components are the **tread**, the **casing**, the **wall** and the **bead**. The wire bead forms the shape and the size of the tyre. The textile plies and the rubber covering runs from one bead to the other. The two main types of tyre construction are radial ply and diagonal ply.

Radial and diagonal ply

Radial ply tyres are so-called because the plies run in a radial manner. The plies of **diagonal ply** tyres usually run at an angle of about 45 degrees to the tyre radius.

Radial ply tyres roll more freely than diagonal ply tyres, which gives better fuel economy and longer tread life. However, diagonal ply tyres are quieter and less inclined to make the suspension or the steering knock or vibrate if a road bump is hit. For this reason diagonal ply tyres are used on some American vehicles — those likely to be used on poorly surfaced roads, or off-road.

Because of their constructional differences, radial ply and diagonal ply tyres behave differently; for this reason they must not be mixed on the same axle. If fitted in a pair, the radial ply tyres must be fitted to the rear wheels.

Nomenclature

The term diagonal ply refers to the fact that the plies run from bead to bead in a diagonal or angular manner. Previously this type of construction was more commonly referred to as cross-ply; this name is still used by older mechanics and tyre fitters.

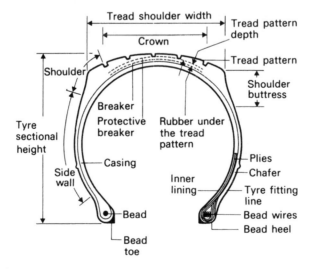

FIGURE 10.5
Parts of a tyre

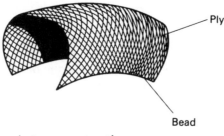

Ply

Bead

Cross-ply tyre construction

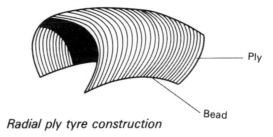

Ply

Bead

Radial ply tyre construction

FIGURE 10.6
Tyre construction

Radial tyre tread for bad weather conditions

Radial tyre tread for high speeds

FIGURE 10.7
Tread patterns

Tyre treads

Different types of tyre treads are used for different purposes. The tyre tread patterns are designed for the different applications of the vehicle. Typical examples are ordinary **multi-purpose** car tyres (budget), **truck tyres**, **high-speed** car tyres, **off-road** vehicle tyres, **asymmetric** treads, **mud and snow** (M&S), **wet race**, **intermediate** and **dry race** – **slicks**.

The purpose of the tyre tread is to **discharge water** and enhance the grip. On average tyres the tread is about 7 mm (1/4 inch) deep when new. M&S tyres need to grip into the soft mud/snow surface, so the tread is almost double the depth of a typical tyre. High speed tyres have treads designed to give the maximum grip and run very **quietly at high speed** with minimum heat generation. Race tyres usually use a much **softer rubber compound** for the tread to give better grip — this means that the tread will not last as long as for road tyres.

Tyre sizes

The size of the tyre depends on the size of the wheel. The **diameter** of the tyre is measured across the bead; it is the same as the diameter of the wheel rim. Most tyre diameters are given in inches (in); increases in size are in one-inch increments. A few specialist tyres are measured in millimetres (mm) — called **metric tyres**; these sizes always correspond to half-inch sizes between the inch increments of regular tyres. The reason for this is so that metric tyres cannot be fitted to regular rims and vice-versa.

The tyre width on radial ply tyres is given in millimetres; on diagonal ply tyres it is given in inches.

For example:

Width Diameter	
165 × 13	*Radial ply*
5.60 × 15	*Diagonal ply*
160 × 313	*Metric radial ply tyre*

Other markings

Between the width and the diameter numbers you will find a double figure number and a pair of letters. The double figure number is the aspect ratio of the tyre, which is the height expressed as a percentage of the width. You will find aspect ratios between about 50 and 80 per cent. Two typical letters that you will find are SR: the S shows that the tyre is suitable for speeds up to 180 kph (112 mph), the R shows that it is a radial ply tyre.

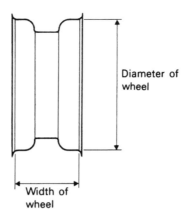

Diameter of wheel

Width of wheel

FIGURE 10.8
Wheel sizes

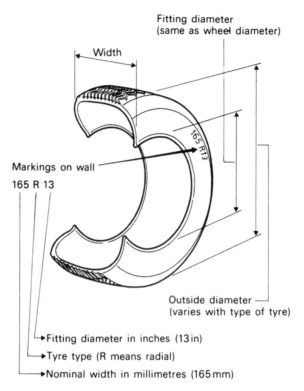

Fitting diameter
(same as wheel diameter)

Width

165 R 13

Markings on wall
165 R 13

Outside diameter
(varies with type of tyre)

Fitting diameter in inches (13 in)

Tyre type (R means radial)

Nominal width in millimetres (165 mm)

FIGURE 10.9
Tyre sizes

For example:

175/65SR13	
175	Width
65	65% aspect ratio
S	Use up to 112 mph
R	Radial ply tyre
13	Wheel diameter

Other information that you will find on the side of a tyre includes country of origin, maximum tyre pressure, maximum load carrying capacity, and compliance with European regulations.

Changing a tyre
JACKING UP TO CHANGE A WHEEL

When jacking up the car to change a wheel the following procedure must be observed:

- place vehicle on level ground
- ensure that the handbrake (parking brake) is on fully

- place chocks behind the wheels that are to remain on the ground
- slacken wheel nuts about a quarter of a turn
- place jack either under the chassis or a suitable suspension member – you should look at the workshop to locate a suitable place.

When refitting the wheel you should use a torque wrench to ensure that it is correctly tightened.

> **RACER NOTE**
>
> Over-tightening the wheel is just as bad as leaving it loose. Over-tightening will also cause loss of time at a tyre change. Always use a torque wrench.

USE OF A TYRE MACHINE

To remove and replace tyres on wheels it is usual to use a tyre machine. The operation of tyre machines varies, and you should read the instruction booklet or have a short training session before operating one.

Tyre pressure

The tyre pressures should be checked regularly. Tyre pressures must be kept within the limits specified by the manufacturer. In either hot or cold weather the tyre pressure may vary. The tyre pressures also vary with the use of the car. The best time to check the tyre pressures is before the car is driven. You should not alter them in the middle of a journey or event unless it is essential for a specific reason.

The use of an accurate tyre gauge is essential; the pencil-type gauges are usually very accurate, for race use digital gauges give the best consistency of readings. Calibrated and certificated gauges are available for use with race and high performance vehicles.

> **RACER NOTE**
>
> Ideal tyre pressures are usually established during practice from lap times and alignment of the cross-tread temperatures.

Tyre tread depth

The tread depth is set by **European regulations**; currently the minimum depth allowed is 1.6 mm. This may be changed at any time.

Because of the inconvenience of changing a wheel after a puncture, and the risk of an accident or indeed closing a traffic lane, you are advised to regularly check all the wheels and tyres, and if a tyre fails to fit a new one.

Tyre wear

If the tyre pressures are set incorrectly the tyres will wear unevenly. If the pressure is **too high** the tread will wear in the **middle**; if the pressure is **too**

low it will wear on the **outer edges**. Faults in the steering and suspension can also cause uneven tyre wear.

Inner tubes

These are circular hollow rubber rings with a valve. The inner tube is used on vehicles fitted with wire-spoked, three-piece and two-piece wheels, to prevent air leakage. The inner tube is inflated inside the wheel and tyre assembly. With wire-spoked wheels a **rim tape** is fitted between the inner tube and the rim, to prevent puncturing by the sharp edges of the spokes. In an emergency, punctures in inner tubes can be repaired with patches; these are larger versions of bicycle patches. However, this practice is not advised and should not be carried out on high-speed vehicles.

Tubeless tyres

Most cars, and a large number of trucks and motorcycles, use tubeless tyres. That is, they do not have inner tubes; the tyre bead forms an **airtight seal against the rim**. Punctures in tubeless tyres can be repaired by inserting a special **rubber plug** into the hole. This must not be done on the tyre sidewall. The plug can be inserted without removing the tyre from the rim. High-speed tyres should only be plugged as a temporary repair and the vehicle should then be driven at a reduced speed.

Tyre valves

Tyre valves hold the air in the inner tube, or they are fitted to the rim in tubeless set-ups. The most popular type is the **Schrader valve** (see Figure 10.10). When air is pumped into the tyre, the valve core is forced downwards to allow the air to pass. When the intake of air is stopped, the pressure of the air in the tyre helps to keep the valve closed.

Tyre damage and repair

Punctures by nails and similar items in the tread of tubeless tyres can be plugged using special plugs and insertion tools without removing the tyre from the wheel. However, this is not recommended for high-performance vehicles, or where high-speed motorway driving is likely.

The maximum length of cut allowed by law is 25 mm (1 inch).

Inner tube punctures can be repaired by patching, but this is not recommended.

Wheel balancing

It is essential to balance the wheels and the tyres as a unit to prevent wheel shimmy. That is, unbalanced wheels may cause the road wheels to shake, which will cause the steering wheel to shake from side to side. Wheel shimmy on out-of-balance wheels is usually noticeable at between 30 and 40 mph.

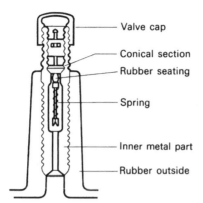

FIGURE 10.10
Schrader valve

FIGURE 10.11
Accessory shop — Caribbean style

Wheel balancing is carried out using a special machine; you will need a short training session to be able to use any specific machine.

Tyre fitting regulations

The laws on tyre fitting and usage in the UK and most of Europe can be summarized as follows:

1. Radial or diagonal ply tyres can be fitted to all vehicles.
2. If only two radial ply tyres are fitted, these must be fitted on the rear wheels.
3. Radial ply and diagonal ply tyres must not be mixed on the same axle.

4. The tyre pressures must be kept within the manufacturer's recommendations.

5. The tread must be not less than 1.6 mm deep for the entire circumference over all the tread width.

6. The tread and the sidewalls must be free from large cuts, abrasions or bubbles.

Wheel rotation

To ensure even wear on all the tyres of a vehicle, it is good practice to move the tyre positions; this should include the spare wheel if possible. It is important to note that all the wheels and tyres must be of the same size and type.

MULTIPLE-CHOICE QUESTIONS

1. The requirements of a wheel are that it must be:
  (a) round and stiff
  (b) round and flat
  (c) stiff and solid
  (d) round and wide

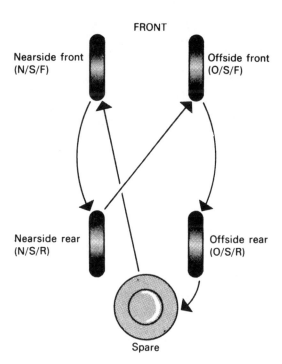

FRONT

Nearside front (N/S/F)

Offside front (O/S/F)

Nearside rear (N/S/R)

Offside rear (O/S/R)

Spare

FIGURE 10.12
Wheel and tyre rotation to give even tyre wear

31 - 2   31-25M01   WHEEL & TYRE

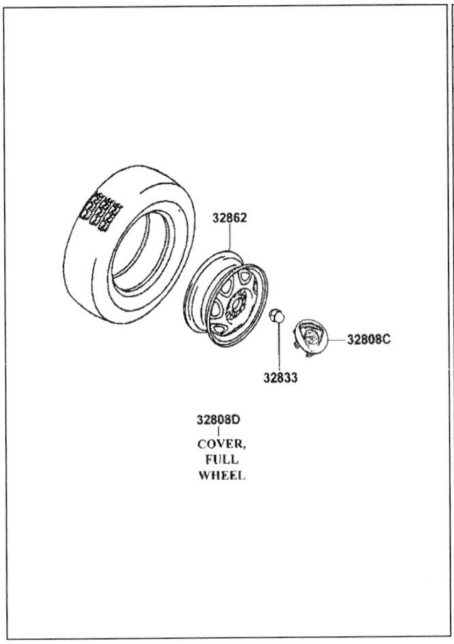

PNC	DESCRIPTION	QTY	REMARKS
32808C	CAP, CENTRE	2	
32808D	COVER, FULL WHEEL	2	
32833	NUT ASSY, HUB	16	STD
23862	WHEEL ASSY	4	15X195 STEEL

32862

32808C

32833

32808D
|
COVER,
FULL
WHEEL

FIGURE 10.13
Wheel and tyre

2. A disadvantage of alloy wheels is that they are:
   (a) brittle
   (b) tough
   (c) wide
   (d) light

3. If only two radial ply tyres are fitted, they must be fitted to:
   (a) both rear wheels
   (b) both front wheels
   (c) opposite corners
   (d) the near side

4. The minimum legal tread depth is:
   (a) 1.6 mm
   (b) 16 mm
   (c) 6 mm
   (d) 0.6 mm

5. One disadvantage of wire wheels is that they are:
   (a) too wide
   (b) too ugly
   (c) easily buckled
   (d) easily cleaned

6. The numbers on a tyre shown in the following sequence — 145/60HR14
   — indicate:
   (a) width, aspect ratio and diameter
   (b) width, pressure and height

(c) load capacity, pressure and diameter
    (d) diameter, price and tread depth
7. For race vehicles it is more consistent to check pressures with:
    (a) a foot pump
    (b) a digital gauge
    (c) an air line
    (d) a pencil gauge
8. Tread, bead and wall are all parts of a:
    (a) tube
    (b) tyre
    (c) rim
    (d) wheel
9. The position of the road wheels on a car is often changed to:
    (a) give even tyre wear
    (b) slow down the tyre wear
    (c) give smooth steering
    (d) give better traction
10. Three-piece rims are found on:
    (a) hatchback cars
    (b) light vans
    (c) trucks
    (d) scooters

(Answers on page 253.)

FURTHER STUDY

1. Make a list of your favourite vehicles and draw up a table to show the diameter, width, aspect ratio, speed rating, tyre type and tyre pressures which are used.
2. Visit a tyre depot and watch how tyres are changed and balanced. You could produce a sketch or a report about your visit.
3. Look at different styles of wheels and sketch your favourite ones, some alloy wheel manufacturers provide CD-ROMs (or other computer software) which allow you to swap pictures of wheels on different cars.

CHAPTER 11

Braking System

KEY POINTS

- Brakes on road vehicles must comply with MOT regulations.
- Brakes work on the principle of converting kinetic energy into heat energy.
- Disc brakes are less prone to brake fade.
- A system of compensation is needed for even braking.
- Hydraulic systems use the principles of Pascal's law.
- Special care must be taken with brake dust and brake fluid.

The main function of the braking system is to stop the vehicle. The braking system also has two less obvious functions: these are to be able to control the speed of the vehicle gradually and gently, and to hold the vehicle when parked on a hill or other incline.

The Highway Code gives a guide to typical **stopping distances** at different speeds. Road and weather conditions affect stopping distances enormously – in wet weather it may take twice as long to stop as in dry weather.

1. MOT REQUIREMENTS

To comply with the requirements of the **Vehicle and Operator Services Agency (VOSA)** the braking system must fulfil these basic minimum requirements:

- Comprise two **independent and separate systems** – usually this is interpreted as the **footbrake** and the **handbrake.**
- **The main braking system** – footbrake – operating on all four wheels, must give a retardation of **50 per cent.**
- **The secondary braking system** – handbrake – operating on only two wheels must give a retardation of **25 per cent** and hold the vehicle in a parked position. This is also referred to as the emergency brake.
- Braking on each axle must be even.

Nomenclature

VOSA is the current name of the part of the UK government which administers the policy on operating vehicles. Other countries have similar bodies, but their names change from time to time – people often just say department of transport. The term MOT, meaning Ministry of Transport (now defunct), is still used on official documents. In the USA, the equivalent department is referred to as DOT.

Vehicles over three years old must pass the **MOT test** each year. One of the tests is that the braking system complies with the minimum requirements (as previously mentioned). Typically, the footbrake must record above 80 per cent retardation and the handbrake 50 per cent. Handbrake travel is usually a maximum of three 'clicks' of the ratchet; footbrake travel is usually about 25 mm (1 inch).

The percentage braking figures are read out from the dials of the **brake testing rolling road**. They relate to **deceleration** (opposite of acceleration) expressed as a **percentage of the acceleration due to gravity** (G). The value for this is 9.81 m/s/s (32 ft/s/s).

> **RACER NOTE**
>
> Brakes on motorsport vehicles need to be perfect in operation – there is no room for error.

2. FRICTION

Friction is the resistance of one body sliding over another body. It is only dependent on the surface finishes of the materials; size of contact area does not affect friction. When two areas are in contact and force is applied to hold them together the friction generates heat.

The braking system uses friction to convert the **kinetic energy** of the vehicle into **heat energy** which is dissipated to the atmosphere.

Nomenclature

Dissipate is a complicated word for dispelling, getting rid of, or spreading about. The heat from the brake pads and discs is dissipated into the atmosphere so that it does not build up and allow the brakes to get hot.

The amount of **heat generated** by the brakes to stop a racing car from say 200 mph (320 kph) will be the same as that generated in the engine to accelerate the car to the same speed.

The amount of friction depends on the materials of the **friction surfaces**, in other words the **pads** and the **discs**, or the **drums** and **shoes**.

The heavier and faster the vehicle, the bigger the brake components will need to be to dissipate the greater amounts of heat generated.

> **RACER NOTE**
>
> Kinetic energy is the energy of motion – the faster a vehicle is travelling and the heavier it is, the more kinetic energy it possesses – so the hotter the brakes will get on a twisty circuit, or a downhill section where the brakes are in constant use.

3. MECHANICAL BRAKES

Mechanically operated brakes are used on vehicles such as **motorcycles** and **karts**. The **handbrakes** on most road cars are also operated mechanically.

A rod or a cable is used to transmit the effort from the lever, or pedal, to the brake shoes. The mechanism used to move the shoes is usually a simple cam arrangement. When the cable, or rod, is pulled the cam is turned so that the brake shoe is pressed against the drum.

Mechanical brakes are very simple, but they are subject to a number of problems, these are:

- cable stretch
- wear of the connecting clevis pins and yokes
- need to adjust each cable run separately.

Compensator

Because it is essential to apply an equal stopping force to each wheel on an axle, whether with mechanical brakes on an old car or kart, or the handbrake on a current vehicle, some form of compensation device is used in the system. That is a mechanical device which distributes the force evenly between the brake assemblies (drum or disc), so that the vehicle

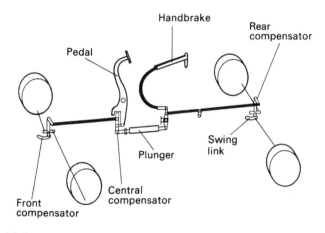

FIGURE 11.1
Cable brake layout — cable brakes are used in karts

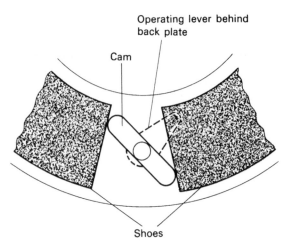

FIGURE 11.2
Cam brake operation

will stop evenly without skidding, or will hold the handbrake evenly on both wheels.

There are two main types of compensator, these are: **swinging link,** also called **swivel tree**, and **balance bar.**

The swinging link has three arms and is mounted on the axle casing. The longitudinal cable pulls on the longer arm, the shorter arms pull at right angles to the longer one that is turning the force through 90 degrees. This changes the direction to transverse and gives a mechanical advantage (leverage) by the difference in arm lengths. It is important to check that the mechanism moves freely − old racing cars may have a grease nipple on the swinging link to encourage lubrication.

The balance beam is used on most current car handbrakes, forming the lower part of the lever. The centre of the beam, which is a very short metal component, is free to twist in the handbrake, the two cables, one to each wheel, are attached to each side of the beam. This allows for slight variations in cable length caused by cable stretch, or uneven adjustment between the two brake assemblies.

4. HYDRAULIC BRAKES

There are many types of hydraulic brake systems, and variation within those systems. The advantage of hydraulic brakes is the hydraulic system is self-compensating − there is no need for mechanical compensation systems and the attendant adjustments to be made; this is explained by Pascal's law.

Pascal's law

Pascal was a French scientist who lived between 1623 and 1662; he spent a lot of time studying why wine bottles break at the bottom when the cork

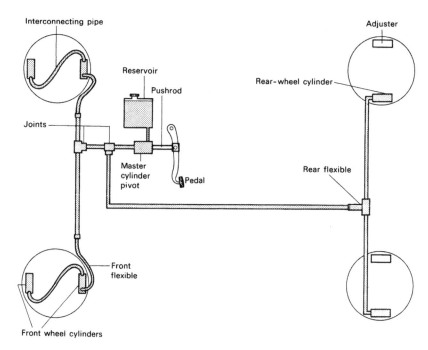

FIGURE 11.3
Single line hydraulic brakes

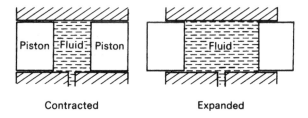

FIGURE 11.4
Double piston wheel cylinder

was pushed in the neck. His discovery was that **a liquid cannot be compressed, and that any pressure applied to a fluid in one direction is transmitted equally in all directions**. This led to the invention of hydraulic brakes about three hundred years later.

Let's have a look at this in more detail. Pascal's theory is applied so that when the driver presses the brake pistons, each of which is double the cross-sectional area of the master cylinder, then the force applied by the driver will be multiplied by eight at the brakes.

Layout of simple system

The simple hydraulic brake system comprises a **master cylinder** and **wheel cylinders** and **drums** — as found on very old racing cars and racing motor

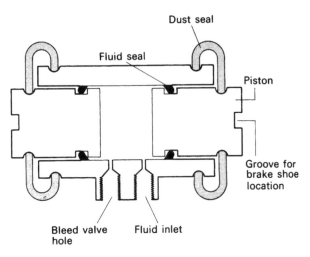

FIGURE 11.5
Twin piston wheel cylinder

cycles, or **callipers** and **discs** (called **rotors** in the US) on later ones. The master cylinder is connected to the wheel cylinders by narrow bore **brake pipes**, also called **brake lines**, or **Bundy tubes**. Remember from Pascal's law — the pressure applied at the master cylinder by pressing the brake pedal will also be applied all along the brake pipes and at the wheel cylinders. When the brakes are fully applied, the pressure is typically 50 bar (750 psi), although it may be as much as 150 bar (2250 psi). When the brakes are released the residual pressure is about 0.25 bar (4 psi). **Flexible brake hoses** are used to connect between the hard lines attached to the body/chassis and the moving components such as the brake callipers.

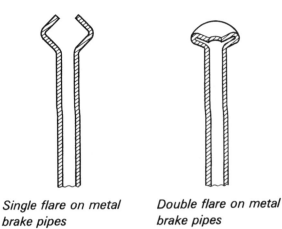

Single flare on metal brake pipes

Double flare on metal brake pipes

FIGURE 11.6
Brake pipe ends

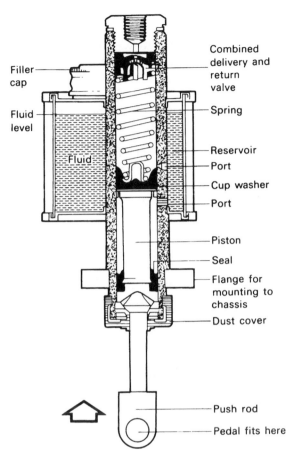

Combined delivery and return valve

Filler cap

Fluid level

Spring

Reservoir
Port
Cup washer
Port

Fluid

Piston

Seal

Flange for mounting to chassis

Dust cover

Push rod

Pedal fits here

FIGURE 11.7
Single piston master cylinder — as used on most single seater cars

Nomenclature

Pipes, hoses, Bundy, lines are all terms that are used and misused in the automotive industry. The same applies to couplings, fitting, connectors, brake nuts and pipe ends. In your assignments and examinations you should try to use the technically correct terms, as always. However, when building a braking system on a car, or bike, you will find reference in the workshop more to the manufacturer, or type — such as Goodridge, AP, and Bembo.

SAFETY NOTE

The pressure of the brake fluid with the pedal depressed is higher than that of the air from the compressor — and you know that must be handled with care — so make sure that the pressure is released before working on the hydraulic system.

5. HANDBRAKE

The handbrake, also called **parking brake**, may be hand operated, or foot operated. On large automatic cars (US standard saloons) the **parking brake is typically applied by the driver's left foot** (the right foot operates both the accelerator and footbrake pedals). The **parking brake is released by a separate hand-operated lever** underneath the dashboard.

On most vehicles the handbrake operates on the two rear wheels; however, some specialist FWD vehicles have the handbrake, on the front wheels to give more emergency stopping power. **Trials cars** have separate handbrakes on each of the two rear wheels; these are called **fiddle brakes** as the drivers fiddle with the two separate levers to maintain traction up steep off-road sections.

RACER NOTE

Sporting trials are run by the **Motor Cycling Club** (MCC) — the oldest motor sport club, which caters for both motorcycles and cars. They use tracks, or off-road sections, where it is necessary to use the fiddle brake to lock one rear driving wheel to maintain grip on the other to climb the hill, or take a very sharp corner by driving one wheel around the other.

The handbrake is held on by a ratchet and pawl mechanism. To check handbrakes for correct operation on a car without using a rolling road, jack up the rear of the car and apply the handbrake one notch (click) at a time. Progressively each wheel should become harder to turn. Both wheels should be fully locked after about three clicks.

The handbrake cables should be visually checked for **fraying** or corrosion — old vehicles may have grease nipples for ease of lubrication; most current handbrake cables have nylon linings to save the need for lubrication — but be

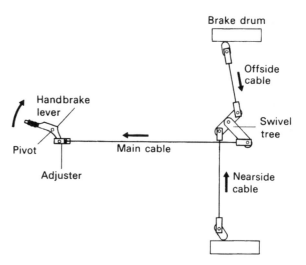

FIGURE 11.8
Handbrake layout — swivel tree type

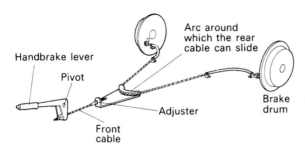

FIGURE 11.9
Handbrake layout (two-cable type)

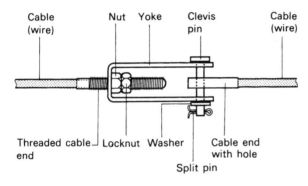

FIGURE 11.10
Handbrake cable adjuster

aware of corrosion, or damage, to the small uncovered section — especially on rally cars where they can become caked in mud. When checking the cables inspect the **clevis pins**, connecting **yokes**, and retaining **split pins** — the application of waterproof grease to the bare cable and linkage is often helpful on rally cars.

6. DRUM BRAKES

Current cars tend to use drum brakes only on the rear with disc brakes at the front.

Master cylinder

This is operated by the driver's right foot. There are two main types, these are:

- **Single piston** type — used on some single-seater cars and older single circuit brakes.
- **Tandem master cylinder**, using two pistons — used on all current dual circuit brake systems.

The operation of both types is very similar. There is a **reservoir** which keeps the main **cylinder** (or chamber) full of fluid. The pedal pushes the **piston** up

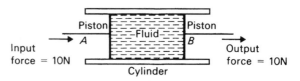

FIGURE 11.11
Equal size pistons in cylinder

$$Pressure = \frac{Force}{Cross\text{-}sectional\ area}$$

the difference in cross-sectional areas. Piston B is twice the area of piston A, so the force on piston B is twice as great.

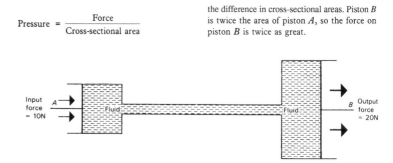

FIGURE 11.12
Unequal size pistons

the cylinder to displace the fluid into the braking system through the **delivery/return valve**.

On a tandem master cylinder there are **two pistons** and **two concentric reservoirs**. Each piston displaces the fluid to its own hydraulic circuit. This means that if one brake line should fracture and leak out the fluid, the vehicle can still be stopped using the remaining parts of the system.

▌ RACER NOTE

In this text you will find reference to older motorsport vehicles — there is more money spent every year, and more races for, historic and classic vehicles than modern ones, and that includes Formula 1. This means that there is more employment in this area of work.

Drum brake shoe layouts

Current vehicle rear drum brakes use **leading and trailing brake shoes**; older race vehicles fitted with drum brakes at the front usually have **twin leading shoes** at the front.

The twin leading shoes give the maximum braking power, the leading edge of the shoe tends to dig into the drum and give what is referred to as a self-servo action. When stopping hard, the front brakes do the bulk of the work — typically the front brakes do 70 per cent of the work compared to 30 per cent at the rear. However, twin leading shoe brakes are only efficient when the

FIGURE 11.13
Always wear a helmet when riding a motorbike

vehicle is going forwards, so they are not useable as a handbrake where the vehicle may be parked on a hill pointing either up, or down. So, for hand-brake operation a leading and trailing shoe arrangement is needed.

Wheel cylinder

There are two main types of wheel cylinders, these are:

- Single piston — used on twin leading shoe brakes where one cylinder operates each shoe.
- Twin piston — used on leading and trailing brakes so that the wheel may be held in either direction of rotation, this is needed to prevent roll back on inclines.

Brake shoes

The **brake linings** are attached to the brake shoes by either rivets or a bonding process (glue). The shoes on ordinary road vehicles are usually fabricated (welded) from plain steel; on race vehicles they are made from aluminium alloy for lighter weight and better heat dissipation. Usually the brake shoes are held in place against the wheel cylinders by strong springs.

RACER NOTE

Before stripping drum brakes it is prudent to either sketch or photograph the layout of springs on shoes.

SAFETY NOTE

Brake shoe springs are very strong, so do not put your fingers where they might get trapped or hurt by the springs slipping.

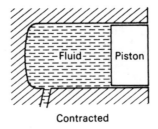

Wheel cylinder (single piston): contracted

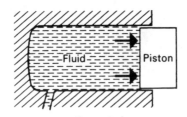

Wheel cylinder (single piston): expanded

FIGURE 11.14
Single piston wheel cylinder

At every major service the brake linings should be checked for wear, the workshop manual will give a minimum thickness figure — typically 3 mm (one-eighth & an inch). If the linings are riveted in place, the rivet heads must be well below the surface of the brake shoe.

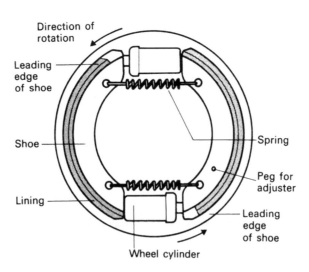

FIGURE 11.15
Twin leading shoe brakes

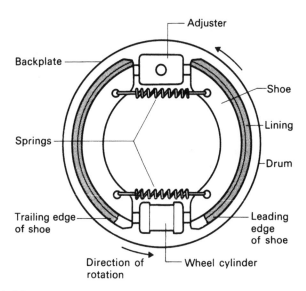

FIGURE 11.16
Leading and trailing shoe brakes

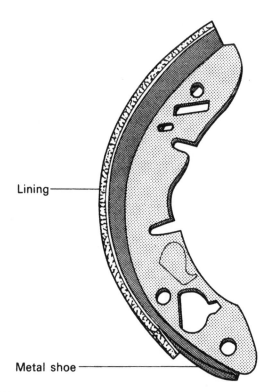

FIGURE 11.17
Brake shoe

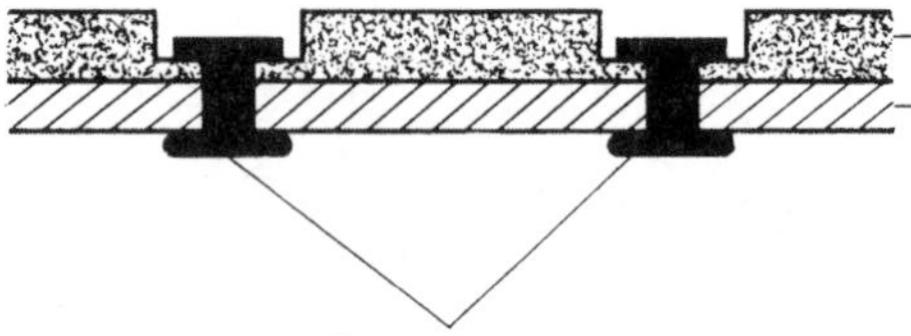

FIGURE 11.18
Brake shoe rivets

Drums

Brake drums may be made from cast iron or aluminium alloy with a steel insert for the friction surface. The aluminium alloy ones are much lighter than the cast iron ones — possibly only a third of the weight. They may also have cooling fins cast on them to help prevent brake fade.

The brake drums are usually held in place on the hub by two small countersunk set-screws. When inspecting the brake pads the drums should be inspected — look for score marks, cracks and ovality.

Nomenclature
Ovality means oval, or in the setting of brake servicing — not properly round. This problem causes brake grab.

7. DISC BRAKES

Disc brakes have several advantages over drum brakes, namely:

- **Less susceptible to brake fade** — that is a reduction in braking efficiency through an increase in the temperature of the friction surfaces, usually after several successive brake applications.
- Open to the air and therefore kept running **cooler**.
- **Easy to change pads**.
- **Greater braking effort** for size and weight with the aid of a brake servo.
- **Self-adjusting**.

On current popular cars it is normal to have front disc brakes with rear drum brakes. High-performance and race vehicles typically have discs both front and rear; the reason for this is that these vehicles usually have an even weight distribution, giving even braking both front and rear.

RACER NOTE

Jaguar's racing reputation was built on winning the Le Mans 24 Hour Race many years ago. They did this by using Dunlop disc brakes to enable them to brake later and harder into each corner, thus increasing their lead on each lap.

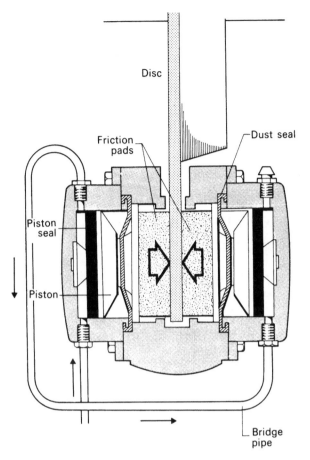

FIGURE 11.19
Disc brake assembly — as used on a classic Jaguar

199

Callipers

Callipers are the disc brake equivalent of the drum brake wheel cylinders. The fluid moves the pistons to press the **pads** against the **disc**. There are many variations of callipers: from single piston callipers used on popular small cars to ones with six pistons used on race and high-performance vehicles.

Nomenclature

You will hear terms such as **four-pot** callipers: this means that each calliper has four pistons — two on each side.

As the pads wear the pistons will expand out to take up the wear. To return the pistons in some cases it is necessary to use a special tool to turn the piston whilst pushing it back into the calliper — look in the workshop manual for the specific application..

Discs (US rotors)

The **pads** act on the discs (rotors) to give the necessary friction. Plain discs are simply cast iron, but most discs are coated with some form of surface finish to improve braking and resist corrosion. The discs are bolted to the hubs.

When overheated, discs can warp, that is like a buckled wheel on a bicycle. This **warping** will give uneven braking and can often be felt at the brake pedal. When replacing pads it is a good idea to check the discs for warping, or as it is called, **run out**.

Special discs

Discs may be manufactured in a number of different ways, mainly to improve cooling, some examples are:

- **Vented discs** – a gap between the two sides of the disc with air vents fitted.
- **Cross drilled discs** – as its name suggests, drilling across the disc.
- **Wavy discs** – a wavy edge to improve give a number of leading edges.

On very advanced race and high-performance cars **carbon discs** are used, these are usually used as multi-floating discs – a system used on large aircraft.

SAFETY NOTE

When replacing brake pads, do one side at a time and to ensure that the seals are not broken and no fluid is lost do not press the brake pedal.

Brake pad thickness

As the brakes are used the **friction material** wears. There is a point at which it becomes necessary to replace the pads – this is when the thickness of the friction material reaches a minimum amount – the vehicle manual will state this figure and how to **measure** it.

Brake pad grade

A variety of different materials are used for brake pads, basically these fit into two major classifications:

- **Soft pads** give more friction, but they generally create more heat and are therefore prone to fade.
- **Hard pads** give less friction, less heat and last longer; hard pads are used for race cars.

Brake lines

Brake lines connect the brake components, transmitting the fluid pressure according to Pascal's law with a small fluid displacement. There are a number of different materials used for brake lines. On road cars they tend to be an

FIGURE 11.20
Pedestrian entrance to a World Championship winner's workshop

alloy steel; on historic racers **copper** is frequently used to look authentic and for its ease of bending; single-seater cars tend to use **stainless steel** for its high value of **hoop stress** – that is its resistance to expand under pressure.

Brake fluid

Brake fluid is a special type of oil developed to give the specific properties needed by the braking system, these are:

- **High boiling point** to reduce the risk of brake fade.
- **Non-corrosive** to rubber and the other materials used in the braking system.
- **Lubricating** properties for the mating parts.
- Will not fail under very **high pressure**.
- **Low viscosity** for rapid response to pedal operation.

When topping up, or replacing, brake fluid ALWAYS USE THE RECOMMENDED BRAKE FLUID – BRAKE FLUIDS MUST NOT BE MIXED.

SAFETY NOTE

Remember, brake fluid removes paint so be careful when you are using it.

RACER NOTE

Hose spanners are available to reduce the risk of damaging brake hose ends.

Brake adjustment

When the friction surfaces wear adjustment is needed to prevent excessive brake pedal travel. There are two main types of adjuster, these are: **wedge** type and **snail cam** type.

201

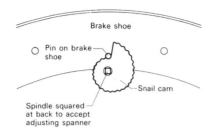
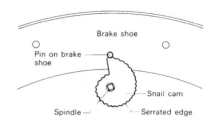

Snail cam adjuster in 'off' position

Snail cam adjuster in 'on' fully adjusted position

FIGURE 11.21
Snail cam adjuster

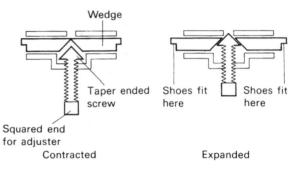

FIGURE 11.22
Taper wedge adjuster

On current cars the rear drum brakes are **self-adjusting.**

FAQs

What are self-adjusting brakes?
There are two main types of self-adjusting brakes: disc bakes are self-adjusting by piston movement in the callipers; rear drum brakes are self-adjusting by using a ratchet mechanism on the handbrake mechanism inside the drum.

Fluid venting

This is also referred to as **bleeding the brakes**. It is the process of **removing air** from the braking system after repairing, or replacing, one of the hydraulic components. There are two main ways of doing this:

- **Manually** – a rubber tube is attached to the **vent nipple** (bleed nipple) by pushing it over the nipple to give a firm fit. The other end of the tube is submerged in a container of clean brake fluid. The nipple is slackened about one turn. When the brake pedal is depressed fluid will flow and bring with it any trapped air. The master cylinder reservoir must be kept topped

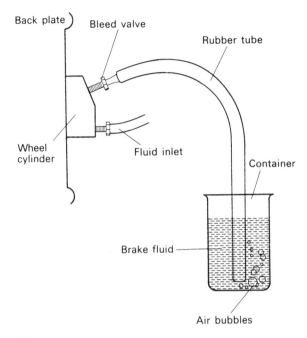

FIGURE 11.23
Brake fluid venting (bleeding) set-up

up throughout this job. When it is seen that air bubbles have stopped flowing, the nipple is retightened and the pedal tested for a firm feel.

- **Using air pressure** — a device is attached to the master cylinder reservoir, which supplies fluid under pressure using air pressure from either a hand pump, or a compressed air supply. The nipple is attached with a tube in the same way as in the manual method. There is no need for anybody to press the pedal. When the nipple is opened fluid will flow — simply close the nipple when the air bubbles stop.

8. SINGLE-SEATER BRAKES

On open wheel, or single-seater, cars it is normal to use **two master cylinders** with a **balance bar** between them to compensate between the front and rear systems. The vehicle in effect has two separate braking systems, one for the front wheel — which does most of the work — and one for the rear wheels. They do not have handbrakes. The pedal pushes on a balance bar, which is a short rod, like a playground see-saw; if the brake push rod is dead centre between the cylinders, the force will be equal at each end. If the balance beam is moved so that the force is more towards one of the master cylinders — this is done by screwing the threaded part of the balance bar in a threaded part of the pedal, there is usually a flexible cable with a knob below the dashboard to allow this to be done easily — the force will be different between front and rear. Adjustments are made to suit the driver and circuit, for instance to prevent rear-wheel lock-up under heavy braking, or to bring the rear round on very tight corners.

35 - 23B00 FRONT CORNER MODULE

PNC	DESCRIPTION	QTY	REMARKS
32300L	KNUCKLE FRONT LH	1	ABS
32300R	KNUCKLE FRONT RH	1	ABS
32502	BLOT, HUB	4	
32502B	BOLT, FR WHEEL HUB	4	12X55, 107
32510A	BEARING KIT	1	
32515A	HUB ASSY, FRONT	1	
32550	SEAL OIL, INNER	1	
32560	SEAL OIL, OUTER	1	
32598	FLANGE NUT, SELF LOCKING	4	

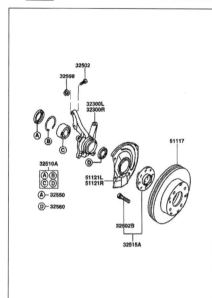

FIGURE 11.24
Front corner module

35-23D00 DUPLEX MODULE

PNC	DESCRIPTION	QTY	REMARKS
54010	MASTER CYLINDER	1	ABS
54011A	RTESERVOIR ASSY	1	
54048	RESERVOIR SEAL	1	
54051	SCREW, BLEEDING	1	
54076	NUT	2	
54093B	SCREW, MOUNTING	2	
54107A	SERVO	1	
54154B	VACUUM HOSE ASSY	1	

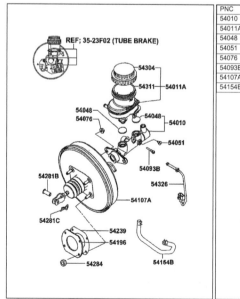

FIGURE 11.25
Duplex module

▌RACER NOTE

Brake fluid is hygroscopic, in other words it attracts water from the atmosphere — it should be replaced at least every season to keep the system fresh.

9. ELECTRONIC CONTROLS

The braking systems on all new road cars have electronic controls such as an **anti-lock braking system** (**ABS**) and traction control. On all-wheel drive (AWD) vehicles these systems can automatically apply the brakes when sensors indicate that an accident is likely, such as when entering a corner too fast, or driving on ice.

MULTIPLE-CHOICE QUESTIONS

1. Braking efficiency can be measured using a:
  (a) rolling road
  (b) steep hill
  (c) dynamometer
  (d) dwell meter

2. Disc brakes on racing cars allow:
  (a) later and harder braking
  (b) softer pedal feel
  (c) better acceleration
  (d) longer braking distances

3. Snail cam and taper wedge are types of:
  (a) disc brakes
  (b) drum brakes
  (c) master cylinder
  (d) brake adjuster

4. Hydraulic brakes are vented (bled) to:
  (a) remove air
  (b) let in air
  (c) remove brake fluid
  (d) top up the brake fluid

5. The leading edge of a brake shoe is the one which:
  (a) touches the brake drum first
  (b) touches the brake drum last
  (c) is used on disc brakes
  (d) is on the right-hand side

6. The wheel cylinder is expanded by:
  (a) brake fluid displacement
  (b) air pressure
  (c) a spring mechanism
  (d) rubber seals

7. Brake lines on racing cars are typically made from:
  (a) iron
  (b) stainless steel

    (c) rubber

    (d) brass

8. Handbrakes are fitted with compensators to allow for:

    (a) broken cables

    (b) uneven adjustment

    (c) ratchet wear

    (d) fluid leaks

9. The footbrake efficiency (expressed as percentage G) for the footbrake must be at least:

    (a) 25 per cent

    (b) 50 per cent

    (c) 75 per cent

    (d) 100 per cent

10. Single-seater cars may use two master cylinders which are adjusted to give brake balance between front and rear to:

    (a) lock the wheels

    (b) enable handbrake turns

    (c) suit the car and circuit

    (d) suit tall drivers

(Answers on page 253.)

FURTHER STUDY

1. Investigate the mechanical brakes used on motorcycles, karts, or bicycles. Some are internal expanding, others are external contracting; see if you can identify the different ones.

2. Carry out a brake test on a vehicle using a rolling road brake tester – do the brakes comply with the MOT requirements?

3. Trace the brake lines and components on a vehicle of your choice and make a black box diagram naming all the components – you may find using a workshop manual is helpful for this.

CHAPTER 12

Electrical and Electronic Systems

KEY POINTS

- The battery is the central store of electricity for the vehicle.
- Battery electrolyte is very corrosive and if splashed on your skin will cause severe injury.
- The alternator is driven by the engine crankshaft to generate electricity to keep the battery charged.
- Most vehicles use the body/chassis as an earth return for the electricity.
- Race cars may use a separate battery on a trolley for starting the engine.
- The engine electronics are controlled by an electronic control unit referred to as the ECU.

SAFETY NOTE

Electrical short circuits are the main cause of vehicle fires; when working on an electrical system it is easy to cause a temporary short-circuit by making two terminals, or cable ends, touch. For this reason you **MUST ALWAYS DISCONNECT THE BATTERY** before carrying out any work on an electrical circuit.

1. BATTERY

The battery is the **central store of electrical energy** for any vehicle. Its purpose is to store electrical energy in chemical form. Most vehicle batteries operate at a nominal 12 volts. Batteries are also available as 6 volt and 24 volt. To control weight distribution, or give added power, or reliability, vehicles

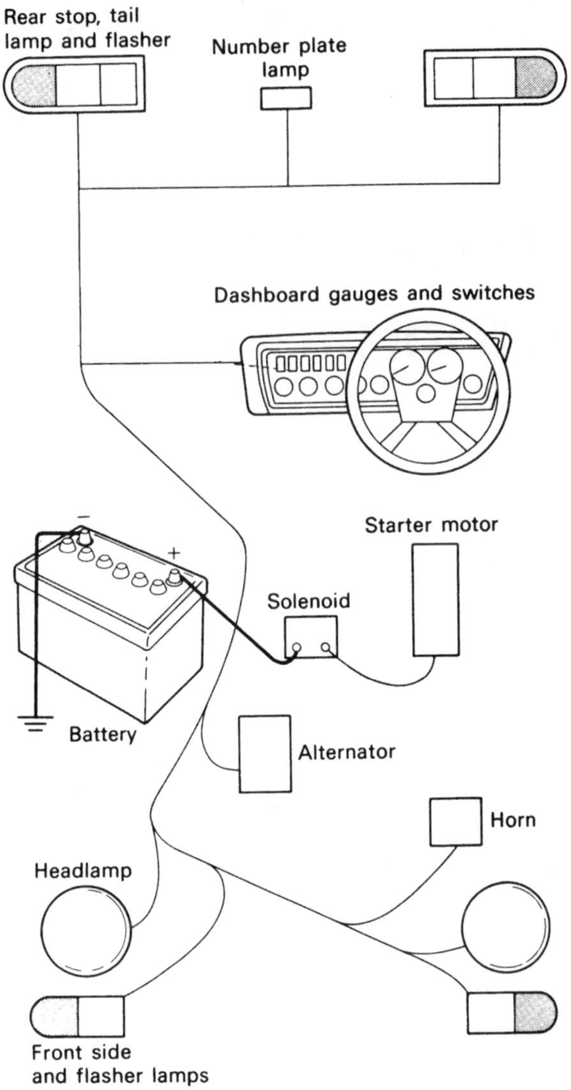

Rear stop, tail
lamp and flasher

Number plate
lamp

Dashboard gauges and switches

Starter motor

Solenoid

Battery

Alternator

Horn

Headlamp

Front side
and flasher lamps

FIGURE 12.1

Layout of electrical components

may use more than one battery. Two 6 volt batteries may be connected in series to give a 12 volt supply; two 12 volt batteries may be connected to give a 24 volt supply. A 12 volt battery may have three terminals to allow both 6 volt and 12 volt supplies.

Nomenclature

The word battery means a group of things and came into use in the seventeenth century as a place where guns and soldiers of the English Civil War were located — in Hampshire

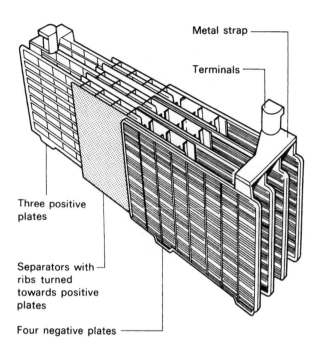

Metal strap

Terminals

Three positive
plates

Separators with
ribs turned
towards positive
plates

Four negative plates

FIGURE 12.2
Battery cell-pack

there is a village called Oliver's Battery. The technically correct name for a car battery is an *accumulator*; you may come across this term when you are working at an advanced level in your studies.

Casing – this is made from a **non-conductive** material which can withstand low-level shock loads, inside it is divided into compartments, or cells.

Cells – each cell is made up of **positive plates** and **negative plates** divided by **porous separators** and submerged in **electrolyte**. There is always an odd number of plates – there is one more negative plate than positive plates to make the maximum use of the positive plates.

The **nominal voltage** of each cell is 2 volts, so six cells are needed to make up a 12 volt battery. We say nominal voltage, as it is not the exact voltage. The voltage given by each cell varies with the state of charge of the cell, and therefore the battery. A fully-charged cell will produce 2.2 volts, meaning that the fully-charged battery will produce 13.2 volts.

Electrolyte – is a solution of **sulphuric acid**; it is **highly corrosive, treat it with care**. Batteries may come **dry charged**, in which case the electrolyte is added only when the battery is purchased – this aids storage and transportation; or it may be **wet charged** – the battery is charged after the electrolyte is added.

The electrolyte in motorsport vehicle batteries is in a **gel** (jelly) form to prevent spillage in the case of an accident.

FIGURE 12.3
Filling a gel battery

SAFETY NOTE

Electrolyte, usually called **battery acid**, is **highly corrosive**; it will burn the skin off your hands and put holes in your overalls. Read the COSHH sheet and take extreme care — **treat batteries with respect.**

Battery charging — batteries are kept fully charged when on the vehicle by the alternator (or dynamo on older vehicles). When not on the vehicle, a battery charger is needed. It is wise to keep a battery fully charged, the battery of a race car, for example, should be fully charged before it is stored for the off-season period. **Maintenance-free batteries** must only be charged at low amperage to avoid damage to the plates; check that the battery charger is the correct type for use with maintenance-free batteries. There are three types of battery charger in use, these are: constant current charging, constant voltage charging and taper charging.

Charging the battery alters the chemical structure in the plates and the electrolyte acidity.

TABLE 12.1 Battery Chemistry		
	Fully charged	**Discharged (flat)**
Positive plate	Lead Peroxide (PbO_2)	Lead Sulphate ($PbSO_4$)
Negative plate	Spongy Lead (Pb)	Lead Sulphate ($PbSO_4$)
Electrolyte	Strong Sulphuric Acid ($2H_2SO_4$)	Dilute (weak) Sulphuric Acid ($2H_2O$ with a percentage of $2H_2SO_4$)

TABLE 12.2 Relative Density Readings			
Colour	**RD or SG**	**State of Charge**	**Comment**
Green	1.280	Fully charged	Leave battery to cool after charging before testing
Yellow or Orange	1.160	Half charged	
Red	1.040	Discharged (flat)	May vary with temperature

Relative density (specific gravity)

Nomenclature

Relative density (RD) is the technically more correct term for specific gravity (SG); in automotive engineering they are both used to mean the same thing. Water at 4 degrees Celsius (39 degree Fahrenheit) as a weight of 1 kg per litre, RD and SG compare the weight of other liquids to this, so sulphuric acid is heavier than water.

The RD or SG of a battery electrolyte is an indication of the state of charge of a battery. This is checked using a hydrometer; this is a glass tube with a float inside it. The denser the liquid, the higher the float will be in the electrolyte. The float may have a numbered scale, or simply a coloured scale to indicate the state of charge.

Maintenance — the following points should be checked:

- **Electrolyte level** — **top up with deionized water** as needed; sealed, maintenance-free batteries should never need this activity.
- **Keep batteries fully charged** — if a vehicle is laid up for the off-season keep it charged with a bench charger.
- **Use proprietary protector** (petroleum jelly or similar) on battery terminals to prevent corrosion.
- **Keep terminals and connectors clean and tightly fastened.**

About half of all breakdowns are caused by battery faults; the AA and the RAC frequently publish statistics on causes of breakdowns.

Starter pack (or **power pack**) — for breakdown and recovery work, a special battery in the shape of a small briefcase (or hand luggage) is used. This is simply charged from the mains with its own cord and plug (230 volt UK; 110 volt Europe, US and most other countries).

Race battery — batteries for race vehicles are very light and usually of the gel type. They do not need to start the vehicle as a portable trolley battery is used for this; that is, a large heavy-duty battery mounted on a trolley. Or a starter pack system may be used, depending on the size of the engine.

RACER NOTE

Keep your trolley battery fully charged and remember which side of the vehicle it plugs into.

211

FIGURE 12.4
Dashboard of the JCB record-holding diesel car

2. ALTERNATOR

The alternator is driven by a 'V' belt from the engine front pulley; it produces **alternating current** (AC) which is converted into **direct current** (DC) to charge the battery. The alternator is made up of a **rotor** which spins inside a **stator** and is encased in adjoining **front and rear casings**. The electricity is generated by the movement of the rotor inside the magnetic stator. The electricity leaves the rotor via two slip rings and brushes. The rear casing houses the **rectifier** and the **regulator**. The rectifier converts the AC current into DC current; the regulator uses a set of **diodes** to control (regulate) the amount of current supplied to charge the battery. Too much electricity would cause the battery to boil and become useless; too little and the battery would become flat. Typically it takes 20 minutes of running the engine to replace the electricity used in starting the car from cold. Much less electrical energy is used starting the engine from warm than starting from cold. Maintenance of the alternator is minimal – check the 'V' **belt tension**, there should be about 12 mm (1/2 inch) free-play on the longest section, **inspect it for cracks or fraying**, and **read the output** using the engine **diagnostic tester**.

Dynamo – before the alternator was the dynamo, this ran at much slower speeds because the commutator's soldered segments would break away above a specific speed – around 6000 rpm.

Magneto – this is used to generate HT for the ignition circuit on small capacity motorcycles (and some historic vehicles). On vehicles, usually motorcycles, where a power supply is also needed, the magneto is combined with a dynamo to make a **mag-dyno**. The mag-dyno is often used on sporting trials motorcycles – the power for the lighting (this is the only use of electricity apart from the ignition) is only provided when the engine is running, and is also dependent on engine speed.

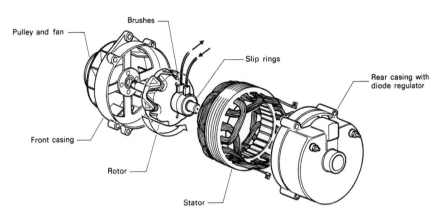

FIGURE 12.5
Components of an alternator

Nomenclature

Alternator, dynamo and mag-dyno are all referred to as *generators* as they all generate electricity.

3. STARTER MOTOR

There are two major types of starter motor, these are:

- **Inertia type** – this uses an inertia drive from the motor to the flywheel so that a mechanical linkage is made only when the starter is turning fast enough.
- **Pre-engaged type** – this has a solenoid to engage the gear on the starter motor with the flywheel ring gear before the starter motor turns.

The actual motor part is similar on both types of starter motor. The **armature** is turned inside the **field coils** of the motor by the effect of supplying a large electrical current to the starter motor brushes. Because of the heavy current involved, a special switch called a **solenoid** is used. This is operated by power

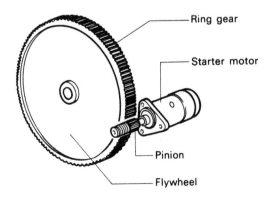

FIGURE 12.6
Ring gear and pinion

supplied to it when the key is turned to the start position. On race cars, a separate starter button may be used, the ignition being controlled by a flick switch with on or off only.

As a guide, a starter motor takes about 180 amps to turn an engine from scratch. Engines typically need to rotate at 120 rpm to initiate combustion; some engines can achieve this in less than half a turn of the crankshaft.

SAFETY NOTE

If you are testing a starter motor when it is off the vehicle, do so in a suitable rig – do not do this loose on the bench top, as the current required could cause a burn, or start an explosion.

4. VEHICLE CIRCUITS

Electrical symbols – to make electrical circuit diagrams easy to understand and small enough to be printed on A4 paper, a system of electrical symbols is used. There are variations between countries and manufacturers, but it is usual for them to be easily identifiable even if you see them for the first time.

Circuit diagrams – most vehicle manufacturers provide circuit diagrams; you should seek out those of vehicles which are of interest to you.

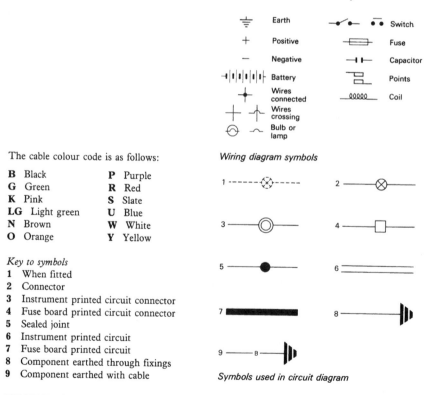

Wiring diagram symbols

The cable colour code is as follows:

B	Black	**P**	Purple
G	Green	**R**	Red
K	Pink	**S**	Slate
LG	Light green	**U**	Blue
N	Brown	**W**	White
O	Orange	**Y**	Yellow

Key to symbols
1 When fitted
2 Connector
3 Instrument printed circuit connector
4 Fuse board printed circuit connector
5 Sealed joint
6 Instrument printed circuit
7 Fuse board printed circuit
8 Component earthed through fixings
9 Component earthed with cable

Symbols used in circuit diagram

FIGURE 12.7

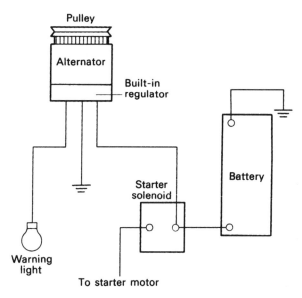

FIGURE 12.8
Alternator wiring diagram

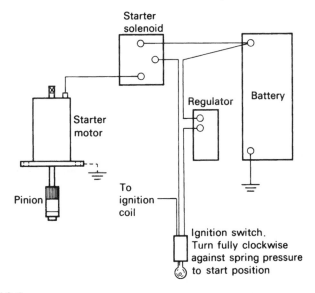

FIGURE 12.9
Starter circuit

▎ RACER NOTE

Beware when using circuit diagrams, they are like the London Underground map: the position of a component on the diagram is not an indication of the position of it on the vehicle.

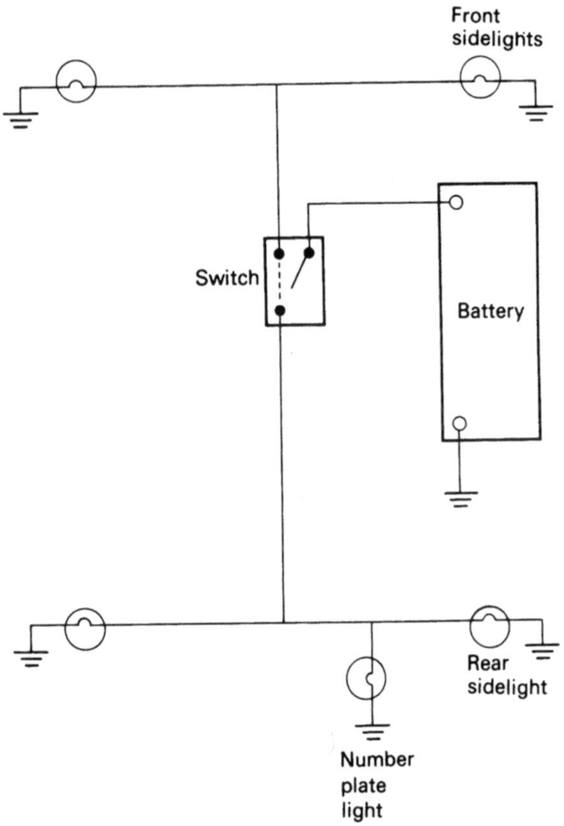

FIGURE 12.10
Sidelight wiring circuit

Chassis earth – this takes large amounts of current, especially when starting the engine (typically 180 amp). It is essential to check all chassis earth points. There is a major lead between the battery and the chassis, and a similar lead between the chassis and the engine.

RACER NOTE

It is prudent to check all earth connections for tightness as part of your spanner check procedure – also use a suitable proprietary anti-corrosion coating to keep them in good condition.

Cables and connectors – all cables on current road cars and motorcycles use crimped ends into semi-locking plastic connectors. The type used varies with the manufacturer. You should investigate those of a manufacturer that interests you.

On older motorsport vehicles, copper wire is mainly used – as compared with the aluminium or gold-plated wires of current high-performance vehicles. Copper wire is usually soldered to brass, or tin-plated connectors on

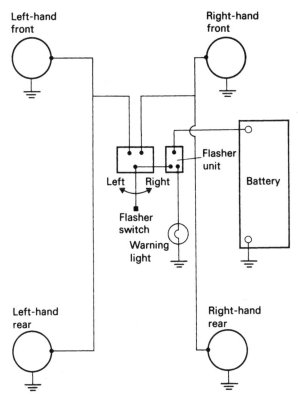

FIGURE 12.11
Flasher circuit

older motorsport vehicles. The soldering is necessary to give good electro-mechanical connection and prevent the wires coming off the terminal.

> **RACER NOTE**
>
> Soldering is a good skill to learn if you intend making up some of your own cable and connector with high electrical and mechanical integrity.

5. LIGHTS

Sidelights are fitted to indicate the size and position of the vehicle to other motorists; in the UK they must be white to the front and red to the rear. Other countries use amber lights at the front. There are regulations as to their position on the vehicle and minimum size. Until recently a 5 watt lamp was compulsory, now LED lights are permitted.

Headlights are fitted to give the driver clear vision in darkness. The law sets regulations on lamp position and beam placement – the rules are devised to prevent them from dazzling other motorists. Headlights may be a number of

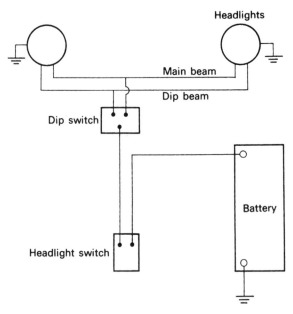

FIGURE 12.12
Headlight wiring circuit

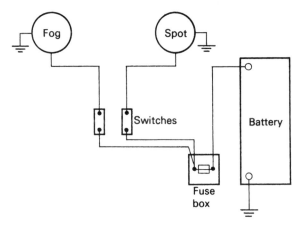

FIGURE 12.13
Fog and spot lamp wiring circuit

different colours, but white is normal, with blue (to look like police cars) the second most popular.

Spot and fog lights — used as their name implies. They should only be used in appropriate conditions.

Direction indicators must show amber in colour, and flash at a rate of 60 flashes per minute; their position is also subject to regulations. Side repeaters may be used too.

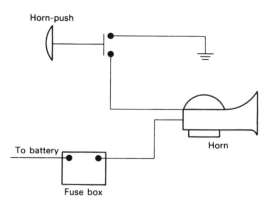

FIGURE 12.14
Horn circuit

FIGURE 12.15
Now that is what you call a sound system

Brake lights are operated by the brake pedal. The minimum is two rear brake lights — 12 watt bulbs, or equivalent LEDs. High level brake lights are normal on current vehicles, with an array of LEDs.

Running lights — these may be the equivalent of sidelights on US vehicles and in some other countries too — they are usually amber in colour.

Race car rear lamps — in bad weather race cars are required to show a red light to the rear; these are almost universally LEDs as they use very little current.

Courtesy — an interior light which illuminates when the doors are opened, or in some cases when the remote locking doors are unlocked.

▌ RACER NOTE
Check operation of ALL lights before taking the vehicle to scrutineering.

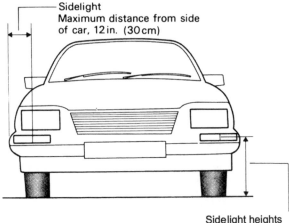

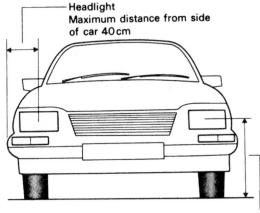

Sidelight
Maximum distance from side
of car, 12 in. (30 cm)

Sidelight heights
Maximum 60 in.
(152 cm)

FIGURE 12.16
Sidelight position

Headlight
Maximum distance from side
of car 40 cm

Headlights height Maximum 42 in. (106 cm)
Minimum 24 in. (61 cm)

FIGURE 12.17
Headlight position

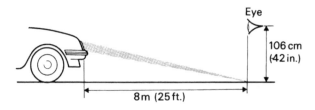

Eye

106 cm
(42 in.)

8 m (25 ft.)

FIGURE 12.18
Headlight alignment

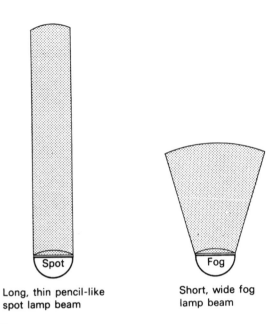

FIGURE 12.19
Spot and fog lamp beams

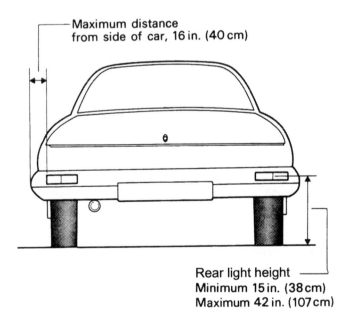

FIGURE 12.20
Rear light position

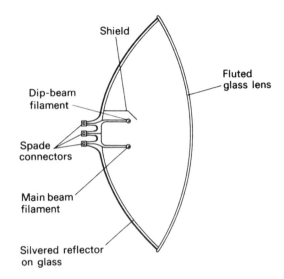

FIGURE 12.21
Headlamp unit

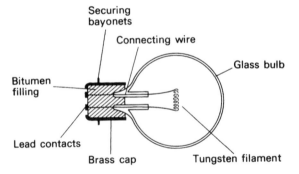

FIGURE 12.22
Twin contact bayonet-type bulb

6. BODY ELECTRICS

Door locks – remote locking (central locking) uses electronically controlled solenoids to operate the mechanical locking mechanism.

Electric windows – these are usually operated by a small electric motor which uses a pulley mechanism to move the window up and down; or turns a cable drive to do the same thing.

Electric mirrors – a small stepper motor moves the mirror within its housing to give accurate adjustment; it is also usual to have a heating element at the back of the mirror glass to clear condensation or frost.

7. ECU

Integrated circuit (IC) – also called a chip. This device performs functions like the SIM card in a mobile phone. The functions are to make decisions based on information from sensors to control the ignition and fuelling.

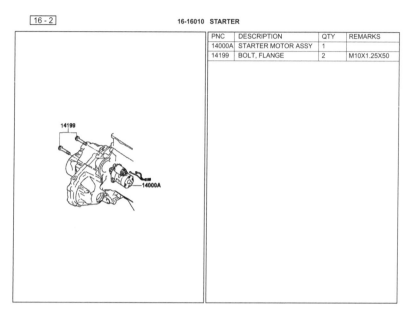

| 16 - 2 | | 16-16010 STARTER | | |

PNC	DESCRIPTION	QTY	REMARKS
14000A	STARTER MOTOR ASSY	1	
14199	BOLT, FLANGE	2	M10X1.25X50

FIGURE 12.23
Starter

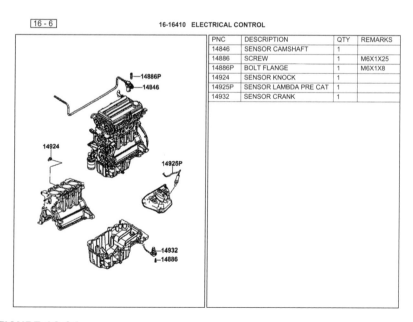

| 16 - 6 | | 16-16410 ELECTRICAL CONTROL | | |

PNC	DESCRIPTION	QTY	REMARKS
14846	SENSOR CAMSHAFT	1	
14886	SCREW	1	M6X1X25
14886P	BOLT FLANGE	1	M6X1X8
14924	SENSOR KNOCK	1	
14925P	SENSOR LAMBDA PRE CAT	1	
14932	SENSOR CRANK	1	

FIGURE 12.24
Electrical control

MULTIPLE-CHOICE QUESTIONS

1. The battery positive plate when fully charged contains:
   (a) lead peroxide
   (b) lead sulphate
   (c) sulphuric acid
   (d) spongy lead

2. The component which keeps the battery fully charged when on the car is the:
   (a) starter motor
   (b) coil
   (c) alternator
   (d) solenoid

3. If you need to top up a lead acid battery you should always use:
   (a) rain water
   (b) deionized water
   (c) tap water
   (d) bottled water

4. A hydrometer reading for a fully charged battery is:
   (a) 1.280
   (b) 1.580
   (c) 1.000
   (d) 1.300

5. The voltage of a fully charged battery cell is:
   (a) 12 volt
   (b) 24 volt
   (c) 2 volt
   (d) 2.2 volt

6. The current drawn by a typical starter motor is:
   (a) 180 amp
   (b) 12 amp
   (c) 24 amp
   (d) 18 amp

7. To maintain both electrical and mechanical integrity when attaching copper wires to connectors you should:
   (a) crimp them
   (b) weld them
   (c) braze them
   (d) solder them

8. Inertia and pre-engaged are both types of:
   (a) generator
   (b) starter motor
   (c) window winder
   (d) battery

9. The type of vehicle which a mag-dyno is likely to be found on is a:
   (a) Formula 1 car
   (b) super bike motorcycle

    (c) sporting trials motorcycle

    (d) single-seater race car

10. The electrical current generated by an alternator is transmitted from the rotor to the rear casing through the:

    (a) slip rings and brushes

    (b) commutator and brushes

    (c) rectifier

    (d) regulator

(Answers on page 253.)

FURTHER STUDY

1. Find any faulty electrical component and open it up to see what is inside it, make notes and draw a diagram to show all the parts.

2. Under the supervision of one of your tutors take voltage, current and resistance readings of a variety of components — note these for future reference.

3. Obtain the circuit diagram for a vehicle of your choice and then draw it out as a large scale plan view as the components appear on the vehicle.

This section defines a number of the words and phrases used in motorsport engineering, including some of the specialist racer and enthusiast vocabulary and jargon

1/4 mile quarter-mile drag racing strip

1/8 mile eighth-mile drag racing strip

24 hour a race lasting 24 hours, the winner is the one covering the greatest distance, usually there are different capacity based classes. For low-budget racers there is an event at Snetterton for Citroen CVs and one in Sussex, near a pub, for lawnmowers

Acceleration rate of increase of velocity

Accessories anything added which is not on a standard vehicle

Ackermann steering set-up to prevent tyre scrub on corners

Add-ons something added after vehicle is made

Adhesion how the vehicle holds the road

Air bags SRS — bags which inflate in an accident

Alignment position of one item against another

Alloy mixture of two or more materials; may refer to aluminium alloy, of an alloy of steel and another metal such as chromium

Ally aluminium alloy

'An off' when you come off the circuit unintentionally into an area where you are not supposed to — if you are lucky it is just grass

Anti-roll bar suspension component to make car stiffer on corners

Atom single particle of an element

Backfire when the engine fires before the inlet valves are closed — sending flames and gas out of the inlet

Barrel cylinder barrel — usually refers to motorcycle engine

BDC bottom dead centre

Bench working surface; also flow bench and test bench

Beta version test version of software, or product

Birdcage tubular chassis frame which resembles a birdcage

Block cylinder block — a number of cylinders in one piece

Block and tackle used to lift engines

Bonnet engine cover (front engined car)

Bore diameter of cylinder barrel

Brooklands first purpose-built racing circuit at Weybridge in Surrey with banking and bridge

Cabriolet four-seater convertible body

Carbon fibre like glass fibre but very strong carbon-based material

Cetane rating resistance to knock of diesel fuel

Chicane sharp pair of bends — often in the middle of a straight

Chocks tapered block put each side of wheel to stop the car rolling

Circuit race circuit

Clerk of Course most senior officer at a motorsport event — person whose decision is final, although there may be a later appeal to the MSA or FIA

Code reader reads fault codes in the ECU of the particular system

Composite material made in two or more layers — usually refers to carbon fibre, may include a honeycomb layer

Compression ratio ratio of combustion chamber size to cylinder bore

Condensation changes from gas to liquid

Con rod connecting rod

Contraction decreases in size

Corrosion there are many different types of corrosion, oxidation/rusting are the most obvious

Cubes cubic inches – American term for size of engine, the saying is, 'there is no substitute for cubes'

Cushion section of seat to sit on

Dashboard instrument panel

Density relative density also called specific gravity

Diagnosis fault finding

Diagnostic machine equipment connected to the vehicle to find faults

Dive vehicle goes down at the front under heavy braking

DOHC double overhead cam

Double 6 another name for a V12 engine

Double 12 a 24 hour event divided into two 12 hours – daytime only with parc firme in the evening

Drag racing two cars racing by accelerating from rest on a narrow drag strip

Engine cover cover over engine (usually refers to rear engined car)

Epoxy resin material used with glass fibre materials

Esses one bend followed by another

Evaporation changes from liquid to gas

Event organizer person who organizes a race or other event

Expansion increases in size

Fast back long sloping rear panels

Fender American for mudguard or wing

FIA Fédération Internationale de l'Automobile – international motorsport governing body

Flag chequered flag, black flag, and red flag

Flag marshal marshal with a flag

Flow bench used to measure the rate of flow of inlet and exhaust gas through a cylinder head

Foam material used to make seats and other items

Force mass × acceleration

Formula car car built for a specific formula, usually refers to open wheeled single-seaters such as Formula Ford or Formula Renault

Friction resistance of one material to slide over another

Gelcoat a resin applied when glass fibre parts are being made – it gives a smooth shiny finish

Glass fibre lightweight mixture of glass material and resin to make vehicle body

GT Grand Turisimo (Italian for Grand Touring) first used on a Ferrari with two passenger doors and a rear luggage opening

Hatchback four-seater with rear upward opening rear

Heat a form of energy

Hill climb individually timed event climbing a hill

Hoist used to lift vehicles, maybe two-post or four-post

Hood American for bonnet

Inboard something mounted on the inside of the drive shafts such as inboard brakes, usually lowers unsprung weight

Inertia resistance to change of state of motion – see Newton's laws

Inertia of motion tendency to keep going

Inertia of rest resistance to changing from stationary to in motion

Intercooler air cooler between turbo charger and inlet manifold – to cool incoming air for maximum density

Kevlar super strong material, often used as a composite with carbon fibre

Le Mans 24 hour race for sports cars, prizes for furthest distance covered and best fuel consumption, organized by Automobile Club de l'Ouest (ACO) (West France Auto Club)

Le Mans Series 24 hour sports car races across the world – many in USA, organized by ACO

Marshal person who helps to control an event

Mass molecular size, for most purpose the same as weight

Metal fatigue metal is worn out

Molecule smallest particle of a material

Monza world's second banked race track in Italy, copy of Brooklands

Motorcross off-road circuit event (motorcycles)

MSA Motor Sports Association

Newton's laws first law – a body continues to maintain its state of rest or of uniform motion unless acted upon by an external unbalanced force

second law – the force on an object is equal to the mass of the object multiplied by its acceleration (F = Ma)

third law – to every action there is an equal and opposite reaction

Nose cone detachable front body section covering front of chassis, may include a foam filler for impact protection

Octane rating resistance to knock of petrol

Off-roader vehicle for going off-road; or an off-road event

OHV overhead valve

Open wheeler race (circuit) car with no wheel covering

'O' rings rubber sealing rings

Original finish original paint work

Outboard something mounted on the outside of the drive shafts such as brakes, usually increases unsprung weight

Oxidation material attacked by oxygen in the atmosphere, for example aluminium turns into a white powdery finish

Paddock where teams are based when not racing

Parc ferme area where competition vehicles are left and can not be worked on or prepared, usually between race rounds or rally stages

Parent metal main metal in an item

Pit place for preparing/repairing/refuelling the vehicle at the side of the circuit

Pit garage garage in the pit lane; also just garage

Pit lane lane off the circuit leading to the pits

Pit wall protecting the pit lane from the circuit

Pot another name for cylinder

Power work done per unit time, HP, BHP, CV, PS, kW

Prepping preparing the vehicle for an event

Projected frontal area area of front of vehicle

Prototype first one made before full production

Rallycross off-road circuit event (rally cars)

Regs racing regulations

Ride height height of vehicle off the road, usually measured from hub centre to edge of wheel arch

Rings piston rings

Roll bar frame inside vehicle which is resistant to bending when vehicle rolls over – safety protection for occupants

Rust oxidation of iron or steel – becomes a reddish colour

Saloon standard four-seater car body

Scrutineer person who checks that a vehicle compiles with the racing regulations, usually when scrutineered the vehicle has a tag or sticker attached

Side valve when the valves are at the side of the engine (old engines and lawnmowers)

Single-seater see formula car

SIPS side impact protection system – door bars (extra bars inside doors)

Skid vehicle goes sideways – without road wheels turning

Speed event any event where cars run individually against the clock

Spine backbone-like structure

Sprinting individually timed event starting from rest over a fixed distance

Sprung weight weight below suspension spring

Squab back of seat – upright part

Squat vehicle goes down at the back under heavy acceleration

Squeal high-pitched noise

Squish movement of air fuel mixture to give better combustion

SRS supplementary restraint system – air bags

Stage rally when the event is broken into a number of individually timed stages, the vehicles start each stage at preset intervals (typically two minutes)

Stall involuntary stopping of engine

Steward a senior officer in the organization of the motorsport event

Straw bails straw bails on side of track for a soft cushion in case of an off

Stress force ÷ cross-sectional area

Strip drag racing strip

Stripping pulling apart

Stroke distance piston moved between TDC and BDC

Swage raised section of body panel

Swage line raised design line on body panel

TDC top dead centre

Temperature degree of hotness or coldness

Test bench test equipment mounted on a base unit

Test hill a hill of which the gradient increases as the top approaches, originally the test was who got the furthest up the hill

Tin top closed car with roof

Torque turning moment about a point (torque = force × radius)

Transporter vehicle to transport competition vehicles to events

Tub race car body/chassis unit made from composite material

Tunnel inverted 'U' section on vehicle floor, on front engined rear-wheel drive cars it houses the propeller shaft – propeller shaft tunnel

Turret American for vehicle roof

Tyre wall wall on side of track built from tyres – giving a soft cushion in case of an off

Unsprung weight weight above suspension spring

Velocity vector quality of change of position, for most purposes the same as speed

Vehicle Maintenance and Repair Level 2

MOTORSPORT ROUTE

TEACHING PROGRAMME

This Institute of the Motor Industry teaching programme sets out the knowledge and understanding necessary for Candidates to complete the **Level 2 Certificate**. However, in planning their delivery, centres should also ensure that Candidates have covered the knowledge and understanding requirements that are set out for Level 1.

The Level 2 programme is broken down into **five sections** including the four main vehicle areas designated by the National Occupational Standards developed by Automotive Skills:

- Chassis Technology
- Engine Technology
- Transmission Technology
- Electrical & Electronic Technology
- General Aspects e.g. health and safety, legal aspects and organizational requirements.

Each centre will have its own approach to the delivery of the programme which will be affected by a number of factors including: Candidates' mode of delivery (part-time day, block release, full-time), Candidates' background and experience, local circumstances and requirements. Centres may wish to integrate or modify the programme to suit their own particular needs and circumstances.

Centres should also note that the total teaching hours shown in this programme are considerably less than the sum total of the individual unit hours. This is because there is often some overlap in content between the individual units. However, when units are grouped together and delivered within a coherent programme these overlaps are removed. If units are delivered on a stand-alone basis the guided learning hours shown on the front of each unit should be noted and the delivery hours adjusted accordingly. Please note that those are only offered as a guide: units may change and emphasis will need to alter to meet the needs of the awarding body and the students.

Chassis Technology

The specific and appropriate health and safety precautions and legislative requirements should be integrated into the delivery of each aspect of chassis technology to ensure relevance and meaning to Candidates. The legislation and generic provisions are outlined in units G1, G2 & G3.

Hours	Topic	Technical Certificate Unit Mapping	On-line Assessment	Practical Assessment
3 hours	Tyre construction: different types of tyre, radial, cross ply, bias belted; tread patterns and uses, competition tyre compounds. Tyre markings: tyre and wheel size markings, speed rating, direction of rotation, profile, load rating, ply rating, tread-wear indicators. Motorsport vehicle wheels: light alloy, pressed steel and wire wheels, flat-edge and double hump rims.	MSM01		
3 hours	Wheel bearing arrangements: non-driving and driven wheel bearing arrangements; fully floating, three-quarter floating and semi-floating. Types of bearing to include: roller, taper roller, needle, ball and plain.	MSM01		
4.5 hours	Steering geometry: castor, camber, kingpin or swivel pin inclination, wheel alignment (tracking), toe-in and toe-out. Ackermann principle, toe-out on turns. Slip angles, neutral steer, oversteer and understeer, slip angles, self-aligning torque.	MSM01		
3 hours	Power-assisted steering: power cylinders, drive belts and pumps, hydraulic valve (rotary, spool and flapper type); introduction to basic principles of electrical and electronic systems, electrical and electronic components.	MSM01		
3 hours	Drum brakes: brake drums, linings and shoes, leading and trailing shoes, self-servo action, automatic adjusters, backing plates, parking brake system, and flies off system.	MSM01		
3 hours	Disc brakes: disc pads, callipers – water cooled, brake disc, ventilated disc, disc pad retraction, parking brake system, electrical and electronic components, wear indicators and warning lamps.	MSM01		

The specific and appropriate health and safety precautions and legislative requirements should be integrated into the delivery of each aspect of chassis technology to ensure relevance and meaning to Candidates. The legislation and generic provisions are outlined in units G1, G2 & G3.—cont'd

Hours	Topic	Technical Certificate Unit Mapping	On-line Assessment	Practical Assessment
3 hours	Hydraulic braking system: single and dual line systems: master cylinders, wheel cylinders, disc brake calliper and pistons, pipes, brake servo, warning lights, parking brakes, equalizing and bias valves.	MSM01		
2 hours	Requirements of brake fluid: properties, hygroscopic action, manufacturer's change periods, fluid classification and rating. Terms associated with mechanical and hydraulic braking systems: braking efficiency, legal requirements, brake fade, brake balance, ABS, EBD.	MSM01		
2 hours	Layout and operation of motorsport vehicle suspension systems: including non-independent − rigid axle, independent front suspension (IFS), independent rear suspension (IRS), leaf spring, coil springs, torsion bar, rubber springs, hydraulic, hydro-pneumatic, hydraulic dampers.	MSM01		
3 hours	Operation of components and types of motorsport vehicle suspension systems: types and components: trailing arms, wishbones, ball joints, track control arms, bump stops, MacPherson strut system, anti-roll bars, stabilizer bars, swinging arms, parallel link, variable, swinging half-axles, transverse link and semi-swinging arms. Introduction to electronic systems.	MSM01		
3.5 hours	Testing procedures for light vehicle chassis systems. Procedures used for inspecting and testing of tyres and wheels, steering, suspension and braking systems including: mechanical system, serviceability, wear, condition, clearances, settings, linkages, joints; hydraulic systems: leaks, adjustments, condition, travel; electrical and electronic systems: security, connections, continuity, earth connections. Use of equipment used for testing of light vehicle chassis systems including: brake	MR01LV, MR04LV, MR05G, G1, G2		

233

(Continued)

The specific and appropriate health and safety precautions and legislative requirements should be integrated into the delivery of each aspect of chassis technology to ensure relevance and meaning to Candidates. The legislation and generic provisions are outlined in units G1, G2 & G3.—cont'd

Hours	Topic	Technical Certificate Unit Mapping	On-line Assessment	Practical Assessment
	roller tester, chassis dynamometer, wheel alignment and steering geometry equipment, manufacturer's dedicated equipment, specialized equipment, electrical testing equipment (multi-meters), pressure gauges. Safety precautions and procedures.			
4.5 hours	Common faults associated with light vehicle chassis systems. Tyres and wheels: abnormal tyre wear, cuts, side wall damage, wheel vibrations, tyre noise (squeal during cornering), tyre over heating (low pressure), tread separation. Steering system: uneven tyre wear, wear on outer edge of tyre, wear on inner edge of tyre, uneven wear, flats on tread, steering vibrations, wear in linkage, damage linkage, incorrect wheel alignment, incorrect steering geometry. Suspension system: wheel hop, ride height (unequal and low), wear, noises under operation, fluid leakage, excessive travel, excessive tyre wear, bounce, poor vehicle handling, worn dampers, worn joints, damaged linkages. Braking systems: worn shoes or pads, worn or scored brake surfaces, abnormal brake noises, brake judder, fluid contamination of brake surfaces, fluid leaks, pulling to one side, poor braking efficiency, lack of servo assistance, brake drag, brake grab, brake fade.	MR01LV, MR04LV, MR05G		
4.5 hours	Procedures for dismantling, removal and replacement of chassis system components. Preparation of tools and equipment used for dismantling, chassis systems and components. Safety precautions, PPE, vehicle protection. Importance of logical and systematic processes.	MR01LV, MR04LV, MR05G, G1, G2		

The specific and appropriate health and safety precautions and legislative requirements should be integrated into the delivery of each aspect of chassis technology to ensure relevance and meaning to Candidates. The legislation and generic provisions are outlined in units G1, G2 & G3.—cont'd

Hours	Topic	Technical Certificate Unit Mapping	On-line Assessment	Practical Assessment
	Inspection and testing of chassis systems and components. Preparation of replacement units for re-fitting. Reasons why replacement components must meet the original specifications (OES). Refitting procedures. Inspection and testing of units and system to ensure compliance with manufacturer's, legal and performance requirements. Inspection and reinstatement of the vehicle following repair to ensure customer satisfaction; cleanliness of vehicle interior and exterior, security of components and fittings, reinstatement of components and fittings. Safety precautions and procedures.			
Total Hours 42	These hours do not include covering Level 1 requirements.			

Engine Technology

The specific and appropriate health and safety precautions and legislative requirements should be integrated into the delivery of each aspect of engine technology to ensure relevance and meaning to Candidates. The legislation and generic provisions are outlined in units G1, G2 & G3.

Hours	Topic	Technical Certificate Unit Mapping	On-line Assessment	Practical Assessment
3 hours	Engine types and configurations: inline, flat, vee, spark ignition, compression ignition, naturally aspirated and turbo-charged engines, hybrid fuel engines. Two- and four-stroke cycles — petrol and diesel. Relative advantages and disadvantages of different engine types and configurations.	MSM02		

(Continued)

The specific and appropriate health and safety precautions and legislative requirements should be integrated into the delivery of each aspect of engine technology to ensure relevance and meaning to Candidates. The legislation and generic provisions are outlined in units G1, G2 & G3.—cont'd

Hours	Topic	Technical Certificate Unit Mapping	On-line Assessment	Practical Assessment
3 hours	Engine components and layouts: crankshafts, pistons, piston rings, connecting rods, flywheel. Single (OHC) and multi-camshaft (DOHC) arrangements. Single and multi-cylinder (two, four, six, eight cylinder) types, in-line, vee.	MSM02		
2 hours	Cylinder head layout and design, combustion chamber and piston design. Calculate compression ratios from given data.	MSM02		
2 hours	Terms related to hydrocarbon fuels: volatility, calorific value, flash point, octane value and cetane value. Composition of hydrocarbon fuels; percentage of hydrogen and carbon in petrol and diesel fuels. Composition of air (percentage of nitrogen, percentage of oxygen).	MSM02		
2 hours	Combustion processes: chemically correct air/fuel ratio for petrol engines 14.7:1 (lambda 1, stoichiometric ratio). Weak and rich air/fuel ratios for petrol engines. Exhaust gas composition and by-products for chemically correct, rich and weak air/fuel ratios of petrol engines: water vapour (H_2O), nitrogen (N), carbon monoxide (CO), carbon dioxide (CO_2), carbon (C), hydrocarbon (HC), oxides of nitrogen (NOx, NO_2, NO) and particulates.	MSM02		
2 hours	Construction and purpose of air filtration systems. Operating principles of air filtration systems. Exhaust system, purpose, layout and design to include brackets, silencers and catalytic converters.	MSM02		
2 hours	Engine lubrication system: splash and pressurized systems, pumps, pressure relief valve, filters, oil ways, oil coolers. Compare wet and dry systems.	MSM02		
2 hours	Terms associated with lubrication and engine oil: full-flow, hydrodynamic, boundary, viscosity, multi-grade, natural and synthetic oil, viscosity index.	MSM02		

The specific and appropriate health and safety precautions and legislative requirements should be integrated into the delivery of each aspect of engine technology to ensure relevance and meaning to Candidates. The legislation and generic provisions are outlined in units G1, G2 & G3.—cont'd

Hours	Topic	Technical Certificate Unit Mapping	On-line Assessment	Practical Assessment
	Requirements and features of engine oil: operating temperatures, pressures, lubricant grades, viscosity, multi-grade oil, additives, detergents, dispersants, anti-oxidants inhibitors, anti-foaming agents, anti-wear, synthetic oils, organic oils and mineral oils.			
2 hours	Layout and construction of internal heater systems, controls and connections within internal heater system. Air conditioning. Basic concept of climate control.	MSM02		
6 hours	Motorsport vehicle diesel fuel systems. Inline and rotary diesel systems, principles and requirements of compression ignition engines, combustion chambers (direct and indirect injection). Function and basic operation of diesel fuel injection components: fuel filters, sedimenters, injectors, injector types (direct and indirect injection), single, multi-hole and pintle nozzle types, governors, fuel pipes, glow plugs, cold start devices, fuel cut-off solenoid. Purpose and basic operation of turbochargers, construction, use of inter-coolers. Procedures for injection pump timing and bleeding the system.	MSM02		
6 hours	Function and layout of petrol injection systems: single and multi-point systems; injection components; injection pump, pump relay, injector valve, air flow sensor, throttle potentiometer, idle speed control valve, coolant sensor, MAP and air temperature sensors, mechanical control devices and electronic control units, fuel pressure regulators, fuel pump relays, lambda exhaust sensors, flywheel and camshaft sensors, air flow sensors (air flow meter and air mass meter) and EGR valve, throttle bodies.	MSM02		
6 hours	Function operation and layout of single and multiple carburettor systems: Weber downdraft and side draft, SU and Stromberg CD types, Mikuni type, Holley four-barrel,	MSM02		

(Continued)

> The specific and appropriate health and safety precautions and legislative requirements should be integrated into the delivery of each aspect of engine technology to ensure relevance and meaning to Candidates. The legislation and generic provisions are outlined in units G1, G2 & G3.—cont'd

Hours	Topic	Technical Certificate Unit Mapping	On-line Assessment	Practical Assessment
	Tillotson pulse diaphragm. Float, fuel and air jets, emulsion tubes, enrichment, idle, bypass, cold start and progression systems, mechanical linkages, electric and mechanical pumps.			
6 hours	Layout of electronic ignition systems advantages over conventional systems (points).	MSM02		
	Electronic ignition circuits and components: LT circuit; battery, ignition switch, electronic trigger devices, capacitor; HT circuit; spark plugs (reach, heat range, electrode features and electrode polarity), rotor arm, distributor (if applicable), distributor cap, ignition leads, ignition coil, ignition timing advance system.			
	Operation electronic system components: amplifiers, triggering systems, inductive pick-ups, Hall generators, optical pulse generators, control units, amplifier units.			
	Ignition terminology: dwell angle, dwell time, dwell variations, advance and retard of ignition timing, ignition timing.			
	Operation of electronic ignition systems under various conditions and loads: engine idling, during acceleration, under full load, cruising, overrun, and cold starting.			
	Basic principles of engine management systems: closed loop system, integrated ignition and injection systems, sensors.			
2 hours	Inspecting engines and engine systems: procedures for mechanical components, induction and air filtration, lubrication system, internal heating system, ignition system, petrol and diesel system, engine management, sensors.	MR02LV, MR05G		
	Procedures to assess serviceability, wear, condition, clearances, settings, linkages, joints, fluid systems, leaks, adjustments; electrical and electronic units; electrical system, operation and functionality, security, connections, continuity, earth connections.			
2 hours	Symptoms and faults associated with engine operation: poor performance, abnormal or	MR02LV		

The specific and appropriate health and safety precautions and legislative requirements should be integrated into the delivery of each aspect of engine technology to ensure relevance and meaning to Candidates. The legislation and generic provisions are outlined in units G1, G2 & G3.—cont'd

Hours	Topic	Technical Certificate Unit Mapping	On-line Assessment	Practical Assessment
	excessive mechanical noise, erratic running, low power, exhaust emissions, abnormal exhaust smoke, unable to start, misfiring, running-on, surging, ignition noise (pinking), excessive fuel consumption, excessive oil consumption, oil leaks, exhaust gas leaks to cooling system, water leaks, water in oil, oil in water, exhaust gas leaks and excessively low or high coolant temperature.			
2 hours	Symptom and faults associated with engine systems: internal heating system, efficiency, operation, leaks, controls; air filtration, air leaks, contamination; lubrication system, low or excessive pressure, oil leaks, oil contamination; diesel fuel system, air in fuel system, water in fuel, filter blockage, leaks, difficult starting, erratic running, excessive smoke (black, blue, white), engine knock, turbocharger faults; petrol injection system, leaks, erratic running, excessive smoke, poor starting, poor performance, poor fuel economy, failure to start, exhaust emissions; ignition system, failure to start hot or cold, erratic running, poor performance, misfire and exhaust emissions.	MR02LV		
6 hours	Procedures for dismantling, removal and replacement of engine units and engine system components: preparation, testing and use of tools and equipment used for dismantling, removal and replacement of engine units and components. Safety precautions, PPE, vehicle protection when dismantling, removal and replacing engine units and components. Logical and systematic processes. Inspection and testing of engine units and components. Preparation of replacement units for refitting or replacement. Reasons why replacement components and units must meet the original specifications (OES) — warranty requirements to maintain performance, safety requirements.	MR02LV, G2		

239

(Continued)

The specific and appropriate health and safety precautions and legislative requirements should be integrated into the delivery of each aspect of engine technology to ensure relevance and meaning to Candidates. The legislation and generic provisions are outlined in units G1, G2 & G3.—cont'd

Hours	Topic	Technical Certificate Unit Mapping	On-line Assessment	Practical Assessment
	Refitting procedures. Inspection and testing of units and system to ensure compliance with manufacturer's, legal and performance requirements. Inspection and reinstatement of the vehicle following repair to ensure customer satisfaction; cleanliness of vehicle interior and exterior, security of components and fittings, reinstatement of components and fittings.			
Total Hours 56	These hours do not include covering Level 1 requirements			

Transmission Technology

The specific and appropriate health and safety precautions and legislative requirements should be integrated into the delivery of each aspect of transmission technology to ensure relevance and meaning to Candidates. The legislation and generic provisions are outlined in units G1, G2 & G3.

Hours	Topic	Technical Certificate Unit Mapping	On-line Assessment	Practical Assessment
4.5 hours	Clutch operating mechanisms: pedal and lever, hydraulic operation, mechanical, cable operated, centrifugal, hydraulic components, master cylinder, slave cylinder, hydraulic pipes, electrical and electronic components (fluid level indicators). Construction and operation of clutches: reasons for fitting a clutch, coil spring clutches, diaphragm spring clutches, single plate clutches, multi-plate clutches, centrifugal clutches, slider clutches.	MSM12		
6 hours	Manual gearboxes: reasons for fitting gearboxes, to provide neutral, reverse, torque multiplication.	MSM12		

The specific and appropriate health and safety precautions and legislative requirements should be integrated into the delivery of each aspect of transmission technology to ensure relevance and meaning to Candidates. The legislation and generic provisions are outlined in units G1, G2 & G3.—cont'd

Hours	Topic	Technical Certificate Unit Mapping	On-line Assessment	Practical Assessment
	Different gearbox types: transverse and inline layouts.			
	Layout and construction of: gears and shafts for four-, five- and six-speed gearbox designs, sliding mesh, constant mesh, dog engagement and synchromesh gearboxes, reverse gear.			
	Construction and operation of gear selection linkages: selector forks and rods, detents and interlock mechanisms.			
	Construction and operation of synchromesh devices.			
	Construction and operation of dog type engagement devices.			
	Arrangements for gearbox bearings: bushes, oil seals, gaskets and gearbox lubrication, speedometer drive.			
	Electrical and electronic components including reverse lamp switch.			
	Calculate gear ratios and driving torque for typical gearbox specifications.			
6 hours	Driveline components: layout and construction of prop-shafts and drive shafts used in front-wheel, rear-wheel and four-wheel drive systems.	MSM12		
	Reasons for using flexible couplings and sliding joints in transmissions systems.			
	Reason for using constant velocity joints in drive shafts incorporating steering mechanisms.			
	Construction and operation of: universal joints, sliding couplings and constant velocity joints.			
	Stresses applied to shafts: torsional, bending and shear.			
	Construction and operation of final drive units: crown wheel and pinion, bevel, hypoid and helical gears; differential gears, sun and planet gears; lubricants, lubrication bearings and seals, limited slip differential, torque biasing differential, spool type final drive, chain drive system, reverse gear for motorcycle engines fitted in cars.			

241

(Continued)

> The specific and appropriate health and safety precautions and legislative requirements should be integrated into the delivery of each aspect of transmission technology to ensure relevance and meaning to Candidates. The legislation and generic provisions are outlined in units G1, G2 & G3.—cont'd

Hours	Topic	Technical Certificate Unit Mapping	On-line Assessment	Practical Assessment
4.5 hours	Reasons for fitting a differential. Calculate final drive gear ratios. Calculate the overall gear ratio from given data (gearbox ratio × final drive ratio). Testing and removal procedures for light vehicle transmission systems: procedures for inspecting and testing clutches and clutch mechanisms including, clearances, pedal and lever settings; cables and linkages; hydraulic system, leaks, adjustments, travel. Procedures used for inspecting and testing gearboxes including, leaks, gear selection, synchromesh operation, abnormal noise. Procedures used for inspecting and testing drive line systems (prop and drive shafts, couplings) including, security, serviceability of rubber boots, leaks, alignment, balance weights (where applicable). Procedures used when inspecting and testing final drive systems including, fluid levels, leaks, noise.	MR12LV, G2		
2 hours	Faults and symptoms associated with transmission systems: clutch faults, gearbox faults, drive line faults (prop-shaft, drive shaft, universal and constant velocity joints), universal joint alignment, final drive faults; faults and symptoms to include mechanical, electrical and hydraulic systems.	MR12LV		
5 hours	Procedures for dismantling, removal and replacement of transmission units: preparation, testing and use of tools and equipment, electrical meters and equipment used for dismantling, removing and replacing transmission systems and components. Safety precautions, PPE, vehicle protection when dismantling, removing and replacing transmission systems and components. Importance of logical and systematic processes. Inspection and testing of transmission systems and components.	MR12LV, G2		

The specific and appropriate health and safety precautions and legislative requirements should be integrated into the delivery of each aspect of transmission technology to ensure relevance and meaning to Candidates. The legislation and generic provisions are outlined in units G1, G2 & G3.—cont'd

Hours	Topic	Technical Certificate Unit Mapping	On-line Assessment	Practical Assessment
	Preparation of replacement units for refitting or replacement of transmission systems or components.			
	Reasons why replacement components and units must meet the original specifications (OES) — warranty requirements, to maintain performance, safety requirements.			
	Refitting procedures.			
	Inspection and testing of units and system to ensure compliance with manufacturer's, legal and performance requirements.			
	Inspection and reinstatement of the vehicle following repair to ensure customer satisfaction; cleanliness of vehicle interior.			
Total Hours 28	These hours do not include covering Level 1 requirements.			

Electrical & Electronic Technology

The specific and appropriate health and safety precautions and legislative requirements should be integrated into the delivery of each aspect of electrical technology to ensure relevance and meaning to Candidates. The legislation and generic provisions are outlined in units G1 & G2.

Hours	Topic	Technical Certificate Unit Mapping	On-line Assessment	Practical Assessment
6 hours	Electrical and electronic principles: Electrical units: volt (electrical pressure), ampere (electrical current), ohm (electrical resistance), watt (power). Requirements for an electrical circuit: battery, cables, switch, current consuming device, continuity. Direction of current flow and electron flow. Series and parallel circuits to include current flow, voltage of components, volt drop,	MR01LV, MR02LV, MR03LV, MR04LV, MR05G, MR12LV		

(Continued)

The specific and appropriate health and safety precautions and legislative requirements should be integrated into the delivery of each aspect of electrical technology to ensure relevance and meaning to Candidates. The legislation and generic provisions are outlined in units G1 & G2.—cont'd

Hours	Topic	Technical Certificate Unit Mapping	On-line Assessment	Practical Assessment
	resistance, the effect on circuit operation of open circuit component(s). Earth and insulated return systems. Cable sizes and colour codes. Vehicle wiring diagrams to include: vehicle lighting, auxiliary circuits, indicators, starting and charging systems. Types of connectors, terminals and circuit protection devices. Electrical and electronic symbols. Meaning of short circuit, open circuit, bad earth, high resistance, electrical capacity. Basic principles of electronic systems and components.			
4.5 hours	Batteries and charging: vehicle batteries including low maintenance and maintenance free; lead acid and nickel cadmium types, cells, separators, plates, electrolyte. Vehicle charging system: alternator, rotor, stator, slip ring, brush assembly, three phase output, diode rectification pack, voltage regulation, phased winding connections, cooling fan, alternator drive systems.	MR01LV, MR02LV, MR03LV, MR05G		
3 hours	Engine starting systems: Inertia and pre-engaged types. Operation of components: inertia and pre-engaged starter motor, starter ring gear, pinion, starter solenoid, ignition/starter switch, starter relay (if appropriate), one-way clutch (pre-engaged starter).	MR01LV, MR02LV, MR03LV, MR05G		
10 hours	Lighting and auxiliary systems: operation of electrical components: front and tail lamps, main and dip beam headlamps, fog and spot lamps, lighting and dip switch, central door locking, anti-theft devices, manual locking and deadlock systems, window winding, demisting systems, door mirror operation mechanisms, interior lights and switching, sunroof operation, directional indicators, circuit relays, bulb types, fan and heater and circuit protection.	MR01LV, MR03LV, MR05G		

The specific and appropriate health and safety precautions and legislative requirements should be integrated into the delivery of each aspect of electrical technology to ensure relevance and meaning to Candidates. The legislation and generic provisions are outlined in units G1 & G2.—cont'd

Hours	Topic	Technical Certificate Unit Mapping	On-line Assessment	Practical Assessment
	Circuit diagrams and operation: side and tail lamps, headlamps, interior lamps, fog and spot lamps, direction indicators central door locking, anti-theft devices, manual locking and deadlock systems, window winding, demisting systems, door mirror mechanisms and sunroof operation. Statutory requirements for vehicle lighting. Headlamp adjustment and beam setting.			
4.5 hours	Testing of electrical and electronic components: safety precautions when working on electrical and electronic systems to include: disconnection and connection of battery, avoidance of short circuits, power surges, prevention of electric shock, protection of electrical and electronic components and circuits from overload or damage. Set-up and use of electrical test equipment: digital and analogue multi-meters, voltmeter, ammeter, ohmmeter, oscilloscope and manufacturer's dedicated test equipment. Electrical and electronic checks: connections, security, functionality, performance to specifications, continuity, open circuit, short circuit, high resistance, volt drop, current consumption and output patterns (oscilloscope).	MR02LV, MR03LV, MR04LV, MR05G, MR12LV, G2		
4.5 hours	Symptoms and faults: high resistance, loose and corroded connections, short circuit, excessive current consumption, open circuit, malfunction, poor performance; battery faults to include flat battery, failure to hold charge, low state of charge, overheating and poor starting.	MR02LV, MR03LV, MR04LV, MR05G, MR12LV, G2		
6 hours	Dismantling, removal and replacement of electrical and electronic components: preparation and use of tools and equipment, electrical meters and equipment used for dismantling, removal and replacement of electrical and electronic systems and components.	MR02LV, MR03LV, MR04LV, MR05G, MR12LV, G2		

245

(Continued)

> The specific and appropriate health and safety precautions and legislative requirements should be integrated into the delivery of each aspect of electrical technology to ensure relevance and meaning to Candidates. The legislation and generic provisions are outlined in units G1 & G2.—cont'd

Hours	Topic	Technical Certificate Unit Mapping	On-line Assessment	Practical Assessment
	Safety precautions, PPE, vehicle protection when dismantling, removal and replacing electrical and electronic components and systems. Logical and systematic processes. Inspection and testing. Preparation of replacement units for refitting or replacement electrical and electronic components and systems. Reasons why replacement components and units must meet the original specifications (OES) – warranty requirements, to maintain performance and safety requirements. Refitting procedures. Inspection and testing of units and system to ensure compliance with manufacturer's, legal and performance requirements. Inspection and reinstatement of the vehicle following repair to ensure customer satisfaction; cleanliness of vehicle interior and exterior, security of components and fittings, reinstatement of components and fittings.			
Total Hours 38.5	These hours do not include covering Level 1 requirements.			

General Aspects

> The specific and relevant health and safety precautions and legislative requirements should be integrated into the delivery of the aspects of vehicle and engine technology wherever possible to ensure relevance and meaning to Candidates; the generic aspects of health and safety are shown below. Only an outline of the main requirements of the Acts and Regulations is required and an awareness of how the legislation relates to the work of trainees.

Hours	Topic	Technical Certificate Unit Mapping	On-line Assessment	Practical Assessment
5 hours	Main Health & Safety legislation: outline main provisions and legal duties imposed by HASAWA, COSHH, EPA, Manual Handling	G1, G2		

The specific and relevant health and safety precautions and legislative requirements should be integrated into the delivery of the aspects of vehicle and engine technology wherever possible to ensure relevance and meaning to Candidates; the generic aspects of health and safety are shown below. Only an outline of the main requirements of the Acts and Regulations is required and an awareness of how the legislation relates to the work of trainees.—cont'd

Hours	Topic	Technical Certificate Unit Mapping	On-line Assessment	Practical Assessment
	Operations Regulations 1992, PPE Regulations 1992. Regulations specific to job role of trainees – awareness of: Health & Safety (Display Screen Equipment) Regulations 1992 Health & Safety (First Aid) Regulations 1981 Health & Safety (Safety Signs and Signals) Regulations 1996 Health & Safety (Consultation with Employees) Regulations 1996 Employers Liability (Compulsory Insurance) Act 1969 Confined Spaces Regulations 1997 Noise at Work Regulations 1989 Electricity at Work Regulations 1989 Electricity (Safety) Regulations 1994 Fire Precautions Act 1971 Reporting of Injuries, Diseases & Dangerous Occurrences Regulations 1985 Pressure Systems Safety Regulations 2000 Waste Management 1991 Dangerous Substances and Explosive Atmospheres Regulations (DSEAR) 2002 Control of Asbestos at Work Regulations 2002 Legislative requirements for use of work equipment: Provision and Use of Work Equipment Regulations 1992 Power Presses Regulations 1992 Pressure Systems & Transportable Gas Containers Regulations 1989 Electricity at Work Regulations 1989 Noise at Work Regulations 1989 Manual Handling Operations Regulations 1992 Health & Safety (Display Screen Equipment) Regulations 1992 Abrasive Wheel Regulations Safe Working Loads Workplace policies and procedures.			

(Continued)

The specific and relevant health and safety precautions and legislative requirements should be integrated into the delivery of the aspects of vehicle and engine technology wherever possible to ensure relevance and meaning to Candidates; the generic aspects of health and safety are shown below. Only an outline of the main requirements of the Acts and Regulations is required and an awareness of how the legislation relates to the work of trainees.—cont'd

Hours	Topic	Technical Certificate Unit Mapping	On-line Assessment	Practical Assessment
6 hours	Organizational requirements for maintenance of the workplace: company health and safety policy, Health and Safety Executive. Trainee's personal responsibilities and limits of their authority with regard to work equipment. Risk assessment of the workplace activities and work equipment. Person responsible for training and maintenance of workplace equipment. When and why safety equipment must be used. Location of safety equipment. Hazards associated with their work area and equipment. Prohibited areas. Plant and machinery that trainees must not use or operate. Why and how faults on unsafe equipment should be reported. Storing tools, equipment and products safely and appropriately. Using the correct PPE. Manufacturers' recommendations. Location of routine maintenance information e.g. electrical safety check log.	G1, G2		
6 hours	Safely use tools and equipment: files, saws, hammers, screwdrivers, drill and drill bits, spanners, punches, measuring equipment, air tools, taps and dies, vices and sockets. Pillar/bench drills, abrasive wheels, presses and lead lights. Axle stands, hydraulic jacks, vehicle lifts/hoists and engine cranes/hoists. Air tools. Welding equipment, power cleaning equipment and exhaust gas extraction. Waste disposal and cleaning equipment. Brake testing equipment. Tool storage and accessibility.	G1		

The specific and relevant health and safety precautions and legislative requirements should be integrated into the delivery of the aspects of vehicle and engine technology wherever possible to ensure relevance and meaning to Candidates; the generic aspects of health and safety are shown below. Only an outline of the main requirements of the Acts and Regulations is required and an awareness of how the legislation relates to the work of trainees.—cont'd

Hours	Topic	Technical Certificate Unit Mapping	On-line Assessment	Practical Assessment
4 hours	Basic maintenance procedures for equipment: Hand tools. Electrical equipment. Mechanical equipment. Pneumatic equipment. Hydraulic equipment. Requirement to clean tools and work area. Requirement to carry out the housekeeping activities safely and to minimize inconvenience to customers and staff. Risks involved when using solvents and detergents. Storage and disposal of waste, used materials and debris correctly. Inspection of equipment to identify faults: Hand tools. Electrical equipment. Mechanical equipment. Pneumatic equipment. Hydraulic equipment. Reporting unserviceable tools and equipment.	G1		
	Economical use of resources: heating, electricity, water, consumables, other energy sources.	G1		
	Requirement to clean tools and work area: maintenance procedures and methodologies, good and safe housekeeping, risks involved with solvents and detergents. Disposal of waste materials: safe system of work, storage and disposal of waste materials, regulations and requirements to dispose of waste, used materials and debris correctly, advantages of recycling.	G1		
3 hours	Requirements when driving vehicles: legal requirements when using vehicles on the road, road safety requirements, lighting, tyres, steering, braking, seat belts, road worthiness.	MR01LV, MR02LV, MR03LV, MR04LV, MR05G, AE06G		

249

(Continued)

The specific and relevant health and safety precautions and legislative requirements should be integrated into the delivery of the aspects of vehicle and engine technology wherever possible to ensure relevance and meaning to Candidates; the generic aspects of health and safety are shown below. Only an outline of the main requirements of the Acts and Regulations is required and an awareness of how the legislation relates to the work of trainees.—cont'd

Hours	Topic	Technical Certificate Unit Mapping	On-line Assessment	Practical Assessment
	Legal requirements for the driver and the vehicle; appropriate drivers licence, road fund licence, vehicle insurance, MOT regulations. Requirements when driving vehicles (company owned, customers') on the road: seat belts, speed limits, care of vehicle, adherence to Highway Code. Requirement of the Road Traffic Act.			
4 hours	Health and safety requirements of vehicle repair: vehicle protection and personal protection (PPE) when working on vehicles. Hazards and risks involved in repair, removal and replacement of units and systems; safety precautions and procedures involved with mechanical, electrical and electronic repair or dismantling. Requirements for disposal of old units, materials, components and fluids. Fire hazards and safety: fire extinguishers, actions in the event of a fire, fire drill and fire exits. Dealing with accidents at work — procedures. Personal conduct in vehicle workshop situations: awareness and care of others and avoidance of inappropriate behaviour.	G2, G3		
6 hours	Technical information relating to vehicle repair: sources technical and repair information: vehicle specifications, identification codes, service schedules, MOT testing requirements, equipment information, procedures for use of equipment repair procedures and test plans. Types of information: paper-based, hard copy manuals, computer stored data, on-board diagnostic displays, CD-ROM, Internet, manufacturer's website. Documentation involved in vehicle repair and maintenance processes: company job cards, manufacturer's service schedules, test plans, inspection sheets, MOT requirements,	G2, G3		

The specific and relevant health and safety precautions and legislative requirements should be integrated into the delivery of the aspects of vehicle and engine technology wherever possible to ensure relevance and meaning to Candidates; the generic aspects of health and safety are shown below. Only an outline of the main requirements of the Acts and Regulations is required and an awareness of how the legislation relates to the work of trainees.—cont'd

Hours	Topic	Technical Certificate Unit Mapping	On-line Assessment	Practical Assessment
	customer requirements and in-vehicle service record. Types of communication: verbal, written, and electronic. Communications involved in vehicle repair. Costs: relationship between time, costs and profit. Economical use of resources, heating, electricity, water, consumable materials e.g. grease. Reporting delays and/or additional work required to relevant supervisory person.			
2 hours	Organization of dealerships and vehicle repair organizations: function of main sections: reception, body shop, repair workshop, paint shop, valeting, parts department, administration office, vehicle sales. Interrelationships of departments. Organizational structure. Job roles.	G3		
2 hours	Developing positive working relationships: Importance. Reasons and effects. Listening to the views of others. Honouring commitments.	G2		
Total Hours 38				

Answers to Multiple-Choice Questions

Question Number	Chapter Number											
	1	2	3	4	5	6	7	8	9	10	11	12
1	d	c	d	b	c	d	b	a	a	a	a	a
2	b	d	d	b	d	c	c	b	b	a	a	c
3	b	a	a	b	d	b	c	c	c	a	d	b
4	c	b	c	b	b	a	d	a	d	a	a	a
5	c	a	b	b	c	b	b	a	c	c	a	d
6	a	d	c	a	b	a	a	b	b	a	a	a
7	a	a	a	b	a	d	b	a	b	b	b	d
8	a	d	a	a	b	a	a	d	a	b	b	b
9	b	b	a	b	b	b	a	c	a	a	b	c
10	c	a	b	c	b	a	d	b	b	d	c	a

Bonnick, Allan. Automotive Computer Controlled Systems, Butterworth-Heinemann (2001)

Chowanietz, Eric. Automobile Electronics, Butterworth-Heinemann (1995)

Hartley, John. The Fundamentals of Motor Vehicle Electrical Systems, Longman (1992)

Livesey, WA. Cassells Motor Vehicle Studies

Livesey, WA. Vehicle Mechanical and Electronic Systems, Institute of the Motor Industry (1996)

Livesey, WA & Robinson, A. The Repair of Vehicle Bodies, Butterworth-Heinemann (2005)

Garrett, TK, Newton, K, and Steeds, W. The Motor Vehicle, Butterworth-Heinemann (2000)

Many of these books are available from the Institute of the Motor Industry on-line book store www.motor.org.uk/store.

Professional motorsport magazines such as *Autosport*, *Track and Race Cars,* and *Motor Sport* are also available.

Index